SpringerBriefs in Applied Sciences and Technology

Manufacturing and Surface Engineering

Series editor

Joao Paulo Davim, Aveiro, Portugal

For further volumes:
http://www.springer.com/series/10623

Santosh S. Hosmani · P. Kuppusami
Rajendra Kumar Goyal

An Introduction to Surface Alloying of Metals

 Springer

Santosh S. Hosmani
Rajendra Kumar Goyal
Department of Metallurgy
    and Materials Science
College of Engineering
Pune
Maharashtra
India

P. Kuppusami
Centre for Nanoscience
    and Nanotechnology
Sathyabama University
Chennai
Tamil Nadu
India

ISSN 2191-530X
ISSN 2191-5318 (electronic)
ISBN 978-81-322-1888-3
ISBN 978-81-322-1889-0 (eBook)
DOI 10.1007/978-81-322-1889-0
Springer New Delhi Heidelberg New York Dordrecht London

Library of Congress Control Number: 2014936433

Printed on acid-free paper

Springer is part of Springer Science+Business Media (www.springer.com)

Preface

There are various engineering applications where surface has to perform a job different from the bulk of a component. On many occasions, just by altering 1–2 % of the total thickness of the components, the properties enhance their performance considerably. In the last several decades, the importance of surface engineering has grown substantially. The list of applications, where manipulation of surface properties is required, is unlimited, especially in the field of automobile, petrochemical, food processing, nuclear, etc. Surface alloying is a class of the surface engineering family, where the surface of the base materials is intentionally alloyed to a thickness of tens of microns. For example, carburizing, nitriding, chromizing, boronizing, etc., are popular methods of surface alloying. These processes involve change in the surface chemistry of the component and modifications in the microstructure and properties. Sometimes, it is advantageous to combine two different surface alloying methods to compensate for the disadvantages offered by one of the methods. The topic of surface alloying is interdisciplinary in nature and various science and engineering streams can work together for its further advancement in science and technology. This book is about the scientific aspects of a focused topic of surface alloying of metals. It is aimed at undergraduate and postgraduate students to develop understanding about the fundamentals and focused topics in the relevant field of surface alloying. This book is just an initiation in this direction and possibly useful to researchers of the R&D institutions and universities, engineers in automobile industries and students who work on this subject for the first time.

The book consists of seven chapters. Chapter 1 (by S.S. Hosmani) deals with the basics of surface alloying. This chapter has attempted to cover the essential concepts of surface alloying along with some of the interesting results in this research area. The relevant concepts are explained by using simple text and schematic diagrams. Chapter 2 (by S.S. Hosmani) attempts to review some of the interesting results associated with nitrided binary iron-based alloys (Fe–Cr and Fe-V alloys). In Chap. 3 (by S.S. Hosmani), the effects of different operating parameters of plasma-nitriding, gas-nitriding, and nitro-carburising of 4330 V steel (NiCrMoV low alloy high strength steel) are discussed. Surface alloying of materials by plasma nitriding using glow discharge plasma has become an important environmentally benign surface modification process to obtain improved hardness, wear resistance, and corrosion resistance. The surface alloying of austenitic stainless steels and Cr-plated austenitic stainless steel by plasma nitriding is described in Chap. 4 (by P. Kuppusami). Chapter 5 (by S.S. Hosmani) is about the role of carburizing, chromizing, and duplex surface treatment on microstructure and hardness of mild steel. Chapter 6 (by R.K. Goyal) attempts to introduce readers to various characterization techniques, which would be useful in the study of surface treated materials. Finally, Chap. 7 (by S.S. Hosmani and P. Kuppusami) gives the conclusions and possible research scope in the field of surface modification.

While preparing this book, we were indebted to several individuals. We are very thankful to them. Many thanks to Springer-India team that provided constant administrative support for the completion of this book.

Santosh S. Hosmani
P. Kuppusami
Rajendra Kumar Goyal

Contents

1 Basics of Surface Alloying 1
  1.1 General Introduction 1
  1.2 Mechanism of Surface Alloying 3
  1.3 Chemical Potential of Surface-Alloying-Atmosphere 3
    1.3.1 Carburizing Potential.............................. 3
    1.3.2 Nitriding Potential................................ 8
  1.4 Equilibrium of Nitriding Atmosphere with Iron Surface 9
  1.5 Diffusion and Case Depth............................. 12
  1.6 Nitrogen Concentration-Depth Profile of Nitrided Iron....... 17
  1.7 Improvement in Mechanical Properties Due
      to Surface Alloying................................. 19
  1.8 Some of the Interesting Researches in Nitriding
      and Carburizing................................... 24
    1.8.1 Nitriding of Fe–Me Alloys and Occurrence
          of Excess Nitrogen 24
    1.8.2 Low-Temperature Carburizing 26
  1.9 Conclusions 27
  References 28

2 Nitriding of Binary Iron-Based Alloys: An Overview. 29
  2.1 Introduction 29
  2.2 Precipitation Morphologies in Nitrided Fe–Me Alloys 30
  2.3 Excess Nitrogen 33
  2.4 Compound-Layer Formation........................... 38
  2.5 Conclusions 39
  References 40

**3 Influence of Process Parameters in Plasma-Nitriding,
  Gas-Nitriding, and Nitro-Carburizing on Microstructure
  and Properties of 4330V Steel.** 43
  3.1 Introduction 44
    3.1.1 Nitriding and Nitro-carburizing of 4330V Steel:
          Motivation and Objectives 45

3.2   Experimental.   46
      3.2.1   Specimen Preparation   46
      3.2.2   Surface Alloying.   46
      3.2.3   Specimen Characterization   51
3.3   Results and Discussion.   51
      3.3.1   Effect of Specimen Geometry.   51
      3.3.2   Nonuniform Iron-Nitride Growth   55
      3.3.3   Effect of Temperature   58
      3.3.4   Effect of Gas Composition.   62
3.4   Comparison of the Results   65
3.5   Conclusions   65
References   66

4   **Recent Advances in Surface Alloying of Austenitic Stainless
    Steel by Plasma Nitriding**.   67
    4.1   Introduction   67
    4.2   Issues in Nitriding of Austenitic Stainless Steels   69
    4.3   Process and Mechanisms   70
    4.4   Surface Alloying by Plasma Nitriding   72
          4.4.1   DC Plasma Nitriding of Austenitic Stainless Steel   72
          4.4.2   Pulsed Plasma Nitriding of Chromium-Plated
                  Austenitic Stainless Steel.   72
          4.4.3   Characterization Techniques.   74
    4.5   Nitriding Behavior of Austenitic Stainless Steels.   75
          4.5.1   Structure, Microstructure, and Microhardness.   75
          4.5.2   Internal Nitriding Model   78
    4.6   Nitriding Behavior of Cr-Plated Austenitic Stainless Steel   80
          4.6.1   Microstructure and Microhardness.   80
          4.6.2   Kinetics of Nitriding   82
          4.6.3   Thermal Stability, Distortion, and Wear Resistance.   83
          4.6.4   Nitriding Mechanism.   84
    4.7   Conclusions   85
    References   86

5   **Chromizing, Carburizing, and Duplex Surface Treatment**   89
    5.1   Introduction   89
    5.2   Experimental.   90
    5.3   Results.   91
    5.4   Discussion   98
    5.5   Conclusions   100
    References   100

6   Characterization of Surface-Treated Materials   103
    6.1   X-Ray Diffraction   103
    6.2   X-Ray Photoelectron Spectroscopy   110
    6.3   Scanning Electron Microscopy   112
    6.4   Transmission Electron Microscopy   116
    6.5   Auger Electron Spectroscopy   118
    6.6   Coating Adhesion   119
    6.7   Measurement of Mechanical Properties   119
          6.7.1   Microindentation Hardness Tester   119
          6.7.2   Nanoindentation Hardness Tester   122
    6.8   Thickness Measurement   123
    References   124

7   Conclusions and Future Scope   125
    7.1   Conclusions   125
    7.2   Future Scope   128
          7.2.1   Effect of Surface Mechanical Attrition Treatment
                  on Surface Alloying of Metals/Alloys   128
          7.2.2   Laser Surface Alloying   128
          7.2.3   Friction-Stir Surface Alloying   129
          7.2.4   Plasma Source Ion Implantation   130
    References   131

About the Book   133

About the Authors

Santosh S. Hosmani is a metallurgist by education. He holds a Bachelor's degree in Metallurgical Engineering from NIT Nagpur and an M.Tech in Process Metallurgy from IIT Bombay. He did his Ph.D. work at Max Planck Institute, Stuttgart, Germany. He completed his Ph.D. in 2006 in the area of nitriding under the guidance of Prof. E. J. Mittemeijer. He continued his research as a post-doctoral researcher in Germany and at CWRU, Cleveland, USA. Since 2009, he has held faculty positions at NIT Surathkal and at IIT Delhi. Currently, he is an Assistant Professor in the Department of Metallurgy and Materials Science, College of Engineering, Pune. He has published several papers in international and national journals and conference proceedings.

P. Kuppusami is currently a Senior Scientist at the Centre for Nanoscience and Nanotechnology, Sathyabama University, Chennai. He holds specializations in the areas of surface modification, thin films, and coatings by plasma and laser methods. In the past he has served as Head, X-ray Diffraction and Surface Engineering Section of the Physical Metallurgy Group at the Indira Gandhi Centre for Atomic Research, Kalpakkam, India. He has worked as a DAAD fellow at Helmholtz Zentrum Berlin, Germany and as a JSPS Fellow at National Institute for Materials Science, Tsukuba, Japan. He has published about 140 papers in peer reviewed journals and conference proceedings. He also has to his credit 170 contributed and invited papers, three book chapters, two review papers, and five patents. He is a life member of the Indian Physics Association and the Indian Institute of Metals.

Rajendra Kumar Goyal received his Ph.D. in Materials Science from IIT Bombay in 2007 and his Bachelor's degree in Metallurgical Engineering from NIT Jaipur in 1996. He joined the Department of Metallurgy and Materials Science, College of Engineering, Pune, India, where he now serves as an Associate Professor. He has industrial and research experience of over 14 years. He has successfully completed several research projects funded by national funding agencies like ISRO and UGC. He has several awards and prizes to his credit. He has published over 40 papers in national and international peer-reviewed journals and has presented more than 60 research papers at conferences. His main research areas are composites/nanocomposites, nanomaterials, characterization of materials, electronic materials, structure-properties relationship, etc.

Chapter 1
Basics of Surface Alloying

Abstract Surface alloying is a widely used method in industries to improve the surface properties of metals/alloys. This chapter is focused on the fundamental scientific aspects of surface alloying of metals. Widely used surface alloying elements involved are interstitial elements such as nitrogen, carbon and substitutional element, chromium, etc. The chapter has attempted to cover the essential concepts of surface alloying along with some of the interesting results in this research area.

Keywords Nitriding · Diffusion · Carburizing · XRD

1.1 General Introduction

Surface alloying is one of the important surface engineering processes. Learning of basic science behind the surface alloying process is essential to appreciate the outcome of the process in technical applications (see also Ref. [1]). In many engineering applications, surface properties have a significant impact on the life of metallic workpieces because the functions that need to be performed by the surface are different from the functions to be performed by the bulk of the workpieces. There are many methods for surface alloying of ferrous alloy components using techniques like, pack, gaseous, plasma, ion beam, and salt-bath. Using the effective surface treatment, less expensive grades of alloys can possibly be used for comparable or even improved service life and performance. Carburizing and nitriding are well-known thermochemical surface treatments to improve the fatigue, tribological, and/or anti-corrosion properties of steel workpieces. There are several surface hardening methods available. One method is to introduce carbon or nitrogen in a workpiece. If strong carbide/nitrides forming alloying elements like Ti, Al, V, Cr, Mo, and/or W are finely dispersed in the matrix, they form carbide/nitride precipitates, which cause large increase in the hardness. Hard surface layer also improves the load bearing capability of the component, for example,

S. S. Hosmani et al., *An Introduction to Surface Alloying of Metals*,
SpringerBriefs in Manufacturing and Surface Engineering,
DOI: 10.1007/978-81-322-1889-0_1, © The Author(s) 2014

austenitic stainless steel. Hard surface layer and ductile core of the material have significantly improved the performance during application.

On carburizing, carbon is incorporated into the surface region of iron-based alloys at usually much high treatment temperatures in the range of 900–1050 °C (within austenitic regime). By contrast, nitriding is performed at temperatures in the range of 450–590 °C, i.e. below the binary eutectoid temperatures (within ferritic regime) of the Fe–N solid solution [2]. Compared to carburizing, negligible changes in the dimensions of the workpieces occur upon nitriding, since the bulk remains ferritic during the treatment.

Nitrided regions can be subdivided into (i) the compound layer adjacent to the surface composed of iron nitrides and (ii) the diffusion zone underneath, where nitrogen is either dissolved or precipitate as alloying element nitrides (Fig. 1.1). The improvement in corrosion and wear properties can be attributed to the compound layer, whereas the diffusion zone improves the fatigue properties, if precipitation of alloying element nitrides takes place.

There are several nitriding methods, e.g., plasma nitriding, salt bath nitriding, and gaseous nitriding. The most well-known method to introduce nitrogen into a (ferritic) workpiece is gaseous nitriding. An eminent advantage of gaseous nitriding is the precise control of the chemical potential of nitrogen in the nitriding atmosphere. At the constant nitriding temperature, by controlling the chemical potential of nitrogen in the atmosphere, it is possible to avoid the formation of compound layer.

Since carburizing is done at high temperature, i.e., within the austenite phase region, chances of formation of cementite (Fe_3C) layer is poor. However, if the chemical potential of carbon in the carburizing atmosphere is very high, cementite formation at the surface can occur. Due to the high temperature, dissociation of such cementite leads to the *metal-dusting* problem: $Fe_3C \rightarrow 3\ Fe_{dust} + C_{graphite}$. At low temperature (i.e., below about 800 °C), pack-carburizing will not be successful due to the inability of solid pack-mixture to generate enough CO gas in the pack (see Sect. 1.3.1). In case of the gas-carburizing, using the controlled atmosphere of CO, CO_2, and N_2 gas mixture, successful carburizing is possible even at lower temperature. The surface region of the carburized iron-based alloy appears similar to that of nitrided alloy. However, the difference is the possibility of formation of cementite layer as a compound layer and the diffusion zone consists of carbides plus dissolved carbon in the surrounding matrix.

Similar to carburizing and nitriding, chromizing is one of the widely used surface alloying technologies to improve the high-temperature oxidation and corrosion resistance of workpiece economically. Various chromizing processes have been developed, for example, pack cementation method, molten-salt technique, and vacuum chromizing process. Pack-cementation is one of the cheapest processes for chromizing. Due to the larger size of the chromium atom than carbon/nitrogen atom, diffusion of chromium is lower than diffusion of carbon in steel at any given temperature of the solid phase. Therefore, chromizing is done at temperatures above 1,000 °C.

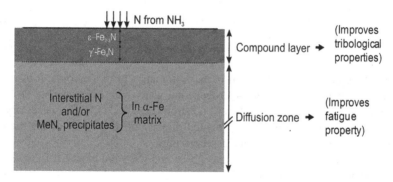

Fig. 1.1 Schematic presentation of the surface region of a nitrided iron/iron-based alloy

1.2 Mechanism of Surface Alloying

The mechanism of surface alloying generally involves *three* steps, which are as follows:

- *Absorption of diffusing species at workpiece surface*: The driving force for this absorption is the difference in chemical potential (or activity) of diffusing species in the surrounding atmosphere ($\mu_{surrounding}$) and at the surface of workpiece ($\mu_{surface}$). At initial stage, absorption of the diffusing species at the surface is high because the difference between $\mu_{surrounding}$ and $\mu_{surface}$ is high. Maximum surface concentration of the species depends on $\mu_{surrounding}$. The absorption of species at the surface generates its concentration gradient.
- *Inward diffusion of the absorbed species*: This causes the transport of species to deeper depths in the cross-section.
- *Formation of compounds*: This depends on the interaction of diffusing species with elements present in the workpiece.

1.3 Chemical Potential of Surface-Alloying-Atmosphere

1.3.1 Carburizing Potential

The ability of carburizing/nitriding atmosphere to introduce carbon/nitrogen into the surface of workpiece depends on the chemical potential of carbon/nitrogen in the atmosphere. The carbon transfer from CO to the solid can occur in principle via the following reactions:

$$2CO \Leftrightarrow [C] + CO_2 \tag{1.1}$$

$$CO + H_2 \Leftrightarrow [C] + H_2O \tag{1.2}$$

where [C] denotes carbon dissolved in the iron matrix or in a carbide. It has been shown experimentally that the heterogeneous water–gas reaction (Eq. (1.2)) is considerably faster than the Boudouard reaction (Eq. (1.1)) [3]. The necessary H_2 for undergoing Eq. (1.2) is either provided directly or as a result of the thermal dissociation of NH_3 (in case of nitrocarburizing).

If local equilibrium between the gas phase and the solid exists, it follows that the carbon activity, a_C, obeys [4, 5]:

$$a_C^S = K^{(1)} r_C^{(1)} \tag{1.3}$$

$$a_C^S = K^{(2)} r_C^{(2)} \tag{1.4}$$

where $K^{(1)}$ and $K^{(2)}$ are the equilibrium constants for Eqs. (1.1) and (1.2), respectively, and $r_C^{(1)}$ and $r_C^{(2)}$ are the corresponding 'carburizing potentials':

$$r_C^{(1)} = \frac{p_{CO}^2}{p_{CO_2}} \tag{1.5}$$

$$r_C^{(2)} = \frac{p_{CO} \cdot p_{H_2}}{p_{H_2O}} \tag{1.6}$$

where p_{CO}, p_{CO_2}, p_{H_2} and p_{H_2O} are the partial pressures of CO, CO_2, H_2 and H_2O gases, respectively.

In the process of pack-carburizing, the workpiece to be carburized is packed in a sealed container. The container can be made of steel or alumina. Workpiece is completely surrounded by granules of charcoal. Pack-carburizing is typically done at 1,000 °C using the pack-mixture of activated charcoal (85 wt%), $BaCO_3$ (10 wt%), $CaCO_3$ (2 wt%), and Na_2CO_3 (3 wt%). However, various compositions of pack-mixture were attempted in the literature [6, 7]. Air trapped in the solid-pack mixture is available for reaction. The charcoal is mixed with activators such as barium carbonate ($BaCO_3$) that promotes the formation of carbon dioxide (i.e. $BaCO_3 \rightarrow BaO + CO_2$). $BaCO_3$ makes CO_2 available at an early stage of carburization and hence it is called "energizer" [6, 7]. CO_2 gas thus generated reacts with carbon in the charcoal to produce carbon monoxide gas (i.e. $CO_2 + C \rightarrow 2CO$). This CO gas reacts with low-carbon steel surface to form atomic carbon that diffuses into the steel. The carbon concentration gradient is generated at the specimen surface which is necessary for diffusion. Diffusion process increases the carbon content to some predetermined depth below the surface to a sufficient level. Elevated carbon content at the specimen surface is necessary to allow subsequent quench hardening.

It must be noted that, even though the carburizing media is in solid state, the actual interaction of the specimen with media is not solid–solid interaction, but the oxygen of the entrapped air (in the carburizing container) initially reacts with carbon of the carburizing medium:

$$C + O_2 \rightarrow CO_2 \qquad (1.7)$$

$$2C + O_2 \rightarrow 2CO \qquad (1.8)$$

As the temperature rises, the following reaction takes place and the equilibrium shifts toward right, i.e., gas becomes progressively richer in CO. At high temperature (>800 °C) the reaction occurs as follows:

$$CO_2 + C \leftrightarrow 2CO \qquad (1.9)$$

This reaction is popularly known as "Boudouard reaction" or "solution loss reaction." At the steel surface the decomposition of CO gas occurs as follows:

$$2C + O_2 \rightarrow CO_2 + [C]_{atomic} \qquad (1.10)$$

$$Fe + [C]_{atomic} \rightarrow Fe(C) \qquad (1.11)$$

where Fe(C) is carbon dissolved in austenite phase near the specimen surface. This atomic and nascent carbon is easily absorbed by the steel surface, and subsequently it diffuses inside the cross-section of the steel specimen. CO_2 thus formed reacts with the carbon (C) of the carburizing medium (reaction 1.9) to produce CO, and thus, the cycle of the reaction continues.

The case depth increases with rise in carburization temperature and time. The best carburizing temperature is 950 °C as the steel surface absorbs carbon at a faster rate and the rate at which it can diffuse inside, thus producing supersaturated case (but, which may produce cracks during quenching) [6, 7]. In pack carburization it is difficult to control the exact case depth because of many factors such as density of packing, amount of air present inside the box, reactivity of carburizer, etc. [7].

Pack-carburizing is difficult at low temperatures. At low temperatures (less than about 680 °C), the formation of CO_2 gas ($C_{(in\ charcoal)} + O_{2\ (in\ trapped\ air)} \leftrightarrow CO_2$) has more negative "Gibbs free energy of equilibrium" (ΔG^0) than that of the formation of CO gas ($2C_{(in\ charcoal)} + O_{2\ (in\ trapped\ air)} \leftrightarrow 2CO$), and the converse is true at high temperatures (more than about 680 °C). The generated CO_2 gas can form CO gas by reacting with carbon in charcoal (reaction 1.9). ΔG^0 and equilibrium constant, K, are given by using the following equations [8]:

$$\Delta G^0 = 171077 - 175.52T \ \left(J\ mol^{-1}\right) \qquad (1.12a)$$

$$\ln(K) = \ln\left(\frac{p_{CO}}{a_{C(solid)} \cdot p_{CO_2}}\right) = -\frac{\Delta G^0}{RT} = -\frac{20577}{T} + 21.111 \qquad (1.12b)$$

Equation (1.12a) is given for temperature range 427–1127 °C. For pure carbon, activity of carbon can be taken as 1. Equation (1.12a) suggests that formation of CO via reaction (1.9) will not occur below 702 °C because ΔG^0 becomes positive. At 800 °C, there can be formation of enough CO–CO$_2$ gases leading to the carburizing potential ($r_C = \frac{p_{CO}^2}{p_{CO_2}}$) of 6.9 (see Eq. (1.12b)). As per the Lehrer diagram (see Ref. [4] and [5]), these carburizing conditions (i.e. $T = 800$ °C; $r_C = 6.9$ atm) should lead to the formation of austenite (solid solution of iron and carbon) at the surface which transforms to the microstructure of ferrite-plus-cementite during slow cooling (see Fig. 1.2) or martensite (see Fig. 1.3) during rapid cooling (quenching) [9]. However, pack-carburization of iron was not successful at or below 800 °C [9]. This discrepancy was due to the following possibilities [9]: (i) the actual temperature of specimen (which was surrounded by the solid pack-mixture) was below the furnace temperature; (ii) the air trapped in the solid pack-mixture was not sufficient to generate enough quantity of carburizing gases at the mentioned temperatures; and (iii) the rate of the gasification reaction decreases rapidly with the decrease in temperature because of the large activation energy barrier [8].

In pack-carburizing, it is difficult to determine the actual carbon potential in the atmosphere surrounding the specimen. However, carburizing and weight measurements (before and after carburizing) of thin foils can be a suitable method to determine the solubility of carbon in the pack-carburized pure iron specimens [9]. It is important to select an appropriate carburizing time so that complete cross-sections of the foil are homogeneously carburized. The carbon uptakes at various temperatures are obtained by weight measurements of the foils [9]. As the temperature increases, the carbon uptake in the specimen enclosed in the carburizing pack-mixture of fixed composition increases. The reason for this observation could be as follows. The solubility of carbon in austenite increases with temperature. Also, as the success of carburizing in pack-mixture depends on the temperature, increase in the carbon uptake could directly be associated with the increase in the carburizing potential with temperature. The amount of carbon content is higher in the quenched specimens than the furnace-cooled specimens [9]. This suggests the possibility of *decarburization* during furnace-cooling of the carburized specimens. Decarburizing could occur due to the loss of carburizing potential in the pack-mixture at low temperatures (say, below about 800 °C, as discussed in the above paragraph).

The gas-carburizing process can be controlled using appropriate flow rates of the gases (e.g., controlled atmosphere of CO/H$_2$/N$_2$ gas mixture). Therefore, carburizing potential can be controlled and successful carburizing can be achieved even at lower temperatures, for example, Gressmann et al. [10] observed the formation of cementite layer on gas-carburized iron at 550 °C. The most important advantage (over pack-carburizing process) of the gas- and plasma- carburizing processes is the possibility of controlling carburizing potential by adjusting the appropriate gas flow rates.

Fig. 1.2 Series of SEM micrographs of the etched cross-section of iron specimen carburized at 950 °C for 2 h (for more details, see Ref. [9])

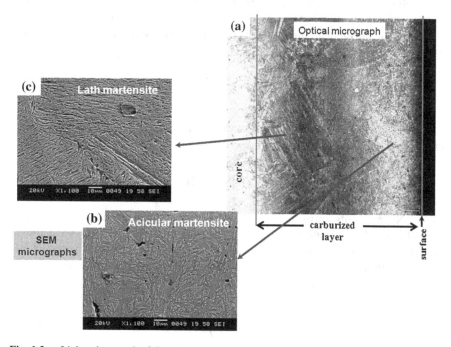

Fig. 1.3 a Light micrograph of the etched cross-section of iron specimen, carburized (at 950 °C for 2 h) and quenched (in ice-cold water); (**b–c**) SEM micrographs of the parts indicated in (**a**), which clearly reveal the acicular (plate like) and lath martensite morphologies near the surface and core, respectively (for more details, see Ref. [9])

1.3.2 Nitriding Potential

Molecular nitrogen is less reactive than ammonia in terms of nitriding of metals. In nitrogen (N_2) atmosphere, the nitriding potential is proportionate to partial pressure of N_2 gas ($[\%N] = k \cdot \sqrt{p_{N_2}}$, where k is equilibrium constant for the following reaction: $1/2N_2 \leftrightarrow [N]_{dissolve\ in\ metal}$). The solubility of nitrogen α-Fe is low (about 0.4 at. %, i.e., 0.12 wt%, at 590 °C). Nitriding of iron using nitrogen gas is impossible because high partial pressure of nitrogen is needed for nitrogen absorption. Covalent bond between N–N atoms is so strong that molecular nitrogen gas will not dissociate into nascent nitrogen at typical nitriding temperature of about 500–600 °C. However, in ammonia (NH_3) environment metals/alloys undergo rapid nitriding reactions. When heated to elevated temperature NH_3 dissociates into N_2 and H_2. Molecular NH_3 should be allowed to dissociate on the steel surface to increase the nitrogen absorption by steel ($[N]_{dissolved\ in\ metal}$). Therefore, continuous supply of fresh, uncracked NH_3 to the steel surface is required. If complete dissociation of NH_3 occurs in the gas atmosphere (before it comes into contact with steel surface), molecular nitrogen forms and this makes nitriding of steel difficult. Temperatures below 600 °C and high flow rate of NH_3 gas can minimize the complete dissociation of NH_3 and, therefore, minimize the formation of molecular nitrogen.

In case of nitriding in ammonia–hydrogen gas mixture, the "nitriding potential," r_N, is defined as follows:

$$r_N = \frac{p_{NH_3}}{p_{H_2}^{3/2}} \tag{1.13}$$

where p_{NH_3} and p_{H_2} are the partial pressures of NH_3 and H_2, respectively. Nitriding in an NH_3–H_2 gas mixture can be presented as [4]:

$$NH_3 \Leftrightarrow [N] + \frac{3}{2}H_2 \tag{1.14}$$

where [N] represents nitrogen dissolved in the ferrite grains of the iron-chromium alloy. Thus, if local equilibrium occurs at the surface of the specimen, the activity of nitrogen in the specimen at its surface, a_N^S, and the chemical potential of nitrogen in the gas phase are both governed by r_N, irrespective of the total pressure of the NH_3–H_2 gas mixture. Thus, it holds for a_N^S:

$$a_N^S = K^{(14)}\left(\frac{p_{NH_3}}{p_{H_2}^{3/2}}\right) = K^{(14)}r_N \tag{1.15}$$

where $K^{(14)}$ is the equilibrium constant for reaction (1.14). Hence, by controlled variation of the gas composition in the nitriding atmosphere, the activity of nitrogen at the surface and thereby the concentration of dissolved nitrogen at the surface can be varied [5].

1.4 Equilibrium of Nitriding Atmosphere with Iron Surface

Gaseous nitriding in controlled atmosphere is close to equilibrium system compared with plasma processes. Therefore, gaseous nitriding is considered here.

Nitriding in NH_3–H_2 gas mixture is equivalent to nitriding in N_2 at a pressure of a number of thousands of atm (thermodynamic argument [4]) and is possible due to the slow thermal decomposition of NH_3 (kinetic argument [5]). Therefore the Fe–N phase diagram established in contact with NH_3–H_2 gas mixture is not the phase diagram for its pure components Fe and N_2 at atmospheric pressure, but represents the phase diagram for Fe and NH_3–H_2 gas mixture.

The chemical potential of nitrogen in a gas phase, $\mu_{N,g}$, consisting of an NH_3–H_2 gas mixture can be defined on the basis of the hypothetical equilibrium:

$$NH_3 \Leftrightarrow \frac{3}{2}H_2 + \frac{1}{2}N_2$$

where

$$\mu_{N,g} \equiv \frac{1}{2}\mu_{N_2} = \mu_{NH_3} - \frac{3}{2}\mu_{H_2}$$

If the standard states refer to unit pressure and if ideal gases or constant fugacity coefficients can be assumed it holds that:

$$\mu_{N,g} = \frac{1}{2}\left[G^0_{N_2} + RT\ \ln p_{N_2}\right] = \left[\left(G^0_{NH_3} + RT\ \ln p_{NH_3}\right) - \left(\frac{3}{2}G^0_{H_2} + RT\ \ln p^{3/2}_{H_2}\right)\right]$$

$$= G^0_{NH_3} - \frac{3}{2}G^0_{H_2} + RT\ \ln\left(\frac{p_{NH_3}}{p^{3/2}_{H_2}}\right)$$

$$= G^0_{NH_3} - \frac{3}{2}G^0_{H_2} + RT\ \ln(r_n)$$

$$(1.16)$$

where p_{N_2}, p_{NH_3} and p_{H_2} are partial pressure of N_2, NH_3 and H_2 respectively; G is Gibbs-free energy per mol, and superscript 0 indicates the standard state; r_n is nitriding potential. From Eq. (1.16), at constant temperature $\mu_{N,g}$ depends only on r_n. If equilibrium is attained between an imposed NH_3–H_2 gas mixture and Fe–N phase (i.e., in present study "say" α-Fe), the chemical potential of nitrogen in α-Fe is equal to that in the gas phase: $\mu_{N,\alpha\text{-Fe}} = \mu_{N,g}$. By variation of the NH_3/H_2 gas mixture (i.e., variation nitriding potential) at a certain temperature and determination of the equilibrium nitrogen content in the α-Fe phase the so-called nitrogen absorption isotherm for Fe in α-phase region can be determined.

A solution of nitrogen in ferrite can be described considering a regular solution of nitrogen on its own sublattice [11]. The nitrogen content on the interstitial sublattice is so low that excess enthalpy need not be taken into account [11]. So now the Gibbs-free energy for one mol Fe–N (α phase) reads as [12]:

$$
\begin{aligned}
G_{\text{Fe}-y_{N,\alpha}} = G_{\text{Fe}}^0 + y_{N,\alpha-\text{Fe}} \cdot G_{N,\alpha-\text{Fe}}^0 \\
+ RT\big[y_{N,\alpha-\text{Fe}} \cdot \ln(y_{N,\alpha-\text{Fe}}) + (1 - y_{N,\alpha-\text{Fe}}) \cdot \ln(1 - y_{N,\alpha-\text{Fe}})\big]
\end{aligned} \quad (1.17)
$$

where $y_{N,\alpha-\text{Fe}}$ is the fraction of sites of "interstitial sublattice" that is occupied by N atoms (i.e., occupancy of nitrogen sublattice for α-Fe phase). A general representation of the chemical potential [5] of nitrogen in ferrite is

$$
\mu_{N,\alpha-\text{Fe}} \equiv N_{Av}\left(\frac{\partial G_{\text{Fe}-y_{N,\alpha}}^{\text{I}}}{\partial y_{N,\alpha-\text{Fe}}}\right) = \left(\frac{\partial G_{\text{Fe}-y_{N,\alpha}}}{\partial y_{N,\alpha-\text{Fe}}}\right) \quad (1.18)
$$

where the symbol G^{I} refers to the Gibbs-free energy of the whole system and symbol G denotes the molar free energy that is independent of the size of the system. N_{Av} is Avogadro's number.

Now from Eqs. (1.17) and (1.18), the chemical potential of nitrogen in ferrite, $\mu_{N,\alpha-\text{Fe}}$, is

$$
\mu_{N,\alpha-\text{Fe}} = G_{N,\alpha-\text{Fe}}^0 + RT \ln\left[\frac{y_{N,\alpha-\text{Fe}}}{1 - y_{N,\alpha-\text{Fe}}}\right]. \quad (1.19)
$$

For equilibrium between NH_3–H_2 gas mixture and ferrite: $\mu_{N,\alpha-\text{Fe}} = \mu_{N,g}$, equating Eqs. (1.16) and (1.19):

$$
\begin{aligned}
G_{N,\alpha-\text{Fe}}^0 + RT \ln\left[\frac{y_{N,\alpha-\text{Fe}}}{1 - y_{N,\alpha-\text{Fe}}}\right] = G_{NH_3}^0 - \frac{3}{2}G_{H_2}^0 + RT \ln(r_n) \\
G_{N,\alpha-\text{Fe}}^0 - G_{NH_3}^0 + \frac{3}{2}G_{H_2}^0 = RT\ln(r_n) - RT\ln\left[\frac{y_{N,\alpha-\text{Fe}}}{1 - y_{N,\alpha-\text{Fe}}}\right].
\end{aligned} \quad (1.20)
$$

Let ΔG^0 be the change in standard Gibbs-free energy for the solubility of nitrogen in ferrite in equilibrium with NH_3–H_2 gas mixture:

$$
NH_3 - \frac{3}{2}H_2 \Leftrightarrow [N]_\alpha. \quad (1.21)
$$

For the above reaction (Eq. (1.21)) ΔG^0 is

$$
\Delta G^0 = G_{N,\alpha-\text{Fe}}^0 - G_{NH_3}^0 + \frac{3}{2}G_{H_2}^0 = -RT \ln\left(\frac{a_n^0}{p_{NH_3}^0 \cdot (p_{H_2}^0)^{-3/2}}\right) \quad (1.22a)
$$

In the standard state, activity of dissolved N in ferrite, $a_n^0 = 1$ and let $[p_{NH_3}^0 \cdot (p_{H_2}^0)^{-3/2}]$ be r_n^0 (where r_n^0 is reference nitriding potential for α-Fe phase). So Eq. (1.22a) becomes

$$\Delta G^0 = G_{N,\alpha-Fe}^0 - G_{NH_3}^0 + \frac{3}{2}G_{H_2}^0 = RT \ln(r_n^0). \tag{1.22b}$$

From Eqs. (1.20) and (1.22b),

$$RT \ln(r_n^0) = RT \ln(r_n) - RT \ln\left[\frac{y_{N,\alpha-Fe}}{1 - y_{N,\alpha-Fe}}\right]$$

$$\left[\frac{y_{N,\alpha-Fe}}{1 - y_{N,\alpha-Fe}}\right] = \left[\frac{r_n}{r_n^0}\right]. \tag{1.23}$$

Since the occupancy of nitrogen for α-Fe phase ($y_{N,\alpha-Fe}$) is very small, the denominator on the left-hand-side of Eq. (1.23) can be replaced by 1:

$$\therefore \quad y_{N,\alpha-Fe} = \frac{r_n}{r_n^0} \tag{1.24}$$

According to Eq. (1.24) the equilibrium nitrogen content in α-Fe at a certain temperature depends linearly on the nitriding potential. Such simple straightforward relations do not occur for other Fe–N phases, like γ'-Fe$_4$N$_{1-x}$ and ε-Fe$_2$N$_{1-z}$ (see Refs. [5, 11] for details).

Putting, $\Delta G^0 = \Delta H^0 - T\Delta S^0$ into Eq. (1.22b) gives

$$RT \ln(r_n^0) = \Delta H^0 - T\Delta S^0 \tag{1.25a}$$

$$\ln(r_n^0) = \left(\frac{-\Delta S^0}{R}\right) + \left(\frac{\Delta H^0}{R}\right)\frac{1}{T} \tag{1.25b}$$

From Eqs. (1.24) and (1.25b)

$$\ln(r_n^0) = -\ln\left(\frac{y_{N,\alpha-Fe}}{r_n}\right) = \left(\frac{-\Delta S^0}{R}\right) + \left(\frac{\Delta H^0}{R}\right)\frac{1}{T} \tag{1.25c}$$

The solubility of nitrogen in ferrite within the temperature range of 300–600 °C is given by [13, 14]:

$$\log\left(\frac{y_{N,\alpha-Fe}}{r_n}\right) = 7.589 - \frac{4025}{T} \tag{1.26}$$

$$\log\left(\frac{y_{N,\alpha-Fe}}{r_n}\right) = 7.395 - \frac{3880}{T} \tag{1.27}$$

where T, $y_{N,\alpha-Fe}$ and r_n are in K, at.% and Pa$^{-1/2}$ respectively.

1.5 Diffusion and Case Depth

Thickness of surface alloyed layer is an important criterion in designing the components in various applications. Layer thickness depends on the rate of transfer of species, i.e., diffusion phenomena.

Diffusion occurs to produce decrease in Gibbs free energy. In practice, it is usually assumed that diffusion occurs down the concentration gradients ("downhill' diffusion). However, this is true under special situations (e.g., spinodal decomposition) and there are some occasions where diffusion can occur up the concentration gradient ("uphill" diffusion). Therefore, the most appropriate explanation for the driving force for diffusion is "chemical potential gradient." As a simple illustration of this consider Fig. 1.4. Two alloys of A-B solid solution, with different compositions (alloy-1 contains X_1 atom fraction of B and the alloy-2 contains X_2 atom fraction of B), are welded together and held at a high temperature. Change in the molar free energy with composition is shown in Fig. 1.4. G_1 and G_2 are the molar free energies of alloys 1 and 2, respectively. Tangents drawn to the free energy curve at X_1 and X_2 give the chemical potentials (μ) of A and B in both alloys. Chemical potential of A is more in alloy-1, whereas the chemical potential of B is more in alloy-2. Therefore, A atoms diffuse from alloy-1 to 2 and B atoms diffuse from alloy-2 to 1. Such diffusion continues till the molar free energy of both alloys decreases to G_3.

There are two widely known mechanisms by which atoms can diffuse through the workpiece: (i) substitutional diffusion (which requires the presence of vacancies) and (ii) interstitial diffusion. Activation energy barrier for substitutional diffusion is larger than the interstitial diffusion. This is because substitutional atoms are larger in size compared with interstitial atoms (e.g., carbon and nitrogen). The substitutional diffusion also requires the presence or creation of vacancies, for example, in the substitutional solid-solution forming alloy A-B, the diffusivity of B is more in the quenched alloy than the diffusivity of B in annealed alloy at a given temperature because the quenched alloy has more concentration of vacancies than the annealed alloy.

The diffusion coefficient (diffusivity), D, increases with temperature. The usual temperature dependence for the diffusion coefficient (D) reads as

$$D = D_0 \cdot \exp\left(\frac{-Q}{RT}\right) \tag{1.28}$$

where D_0 is pre-exponential factor; Q is the activation energy for diffusion, and R is the universal gas constant (=8.314 J mol^{-1} K^{-1}). An Arrhenius plot gives the linear dependence of logarithm of D with inverse of T, where slope of the line gives Q and intercept on Y-axis gives D_0.

Diffusion coefficient (diffusivity) of carbon in ferrite (α-Fe) and austenite (γ-Fe) is given by the following expressions [15]:

Fig. 1.4 Free energy and chemical potential changes during diffusion

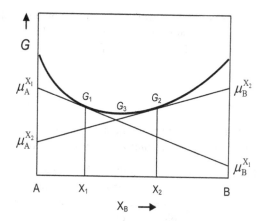

$$D_C^{\alpha-Fe} = \left(6.2 \times 10^{-7}\right) \cdot \exp\left(\frac{-80000}{RT}\right) \quad m^2 sec^{-1} \tag{1.29}$$

$$D_C^{\gamma-Fe} = \left(2.3 \times 10^{-5}\right) \cdot \exp\left(\frac{-148000}{RT}\right) \quad m^2 sec^{-1} \tag{1.30}$$

At a given temperature, diffusion of carbon in austenite (face centered cubic iron) is more difficult than in ferrite (body centered cubic iron), as indicated by the higher activation energy barrier for diffusion of carbon in γ-Fe than in α-Fe, because f.c.c. crystal structure closely packed than b.c.c. structure (atomic packing efficient of 74 % for f.c.c. versus 68 % for b.c.c.).

Diffusion coefficient of nitrogen in ferrite is given by the following expressions [16]:

$$D_N^{\alpha-Fe} = \left(6.6 \times 10^{-7}\right) \cdot \exp\left(\frac{-77822}{RT}\right) \quad m^2 sec^{-1} \tag{1.31}$$

From Eqs. (1.29) and (1.31), the diffusivity of nitrogen in ferrite is slightly easier than that of carbon. Diffusivity of nitrogen in ferrite of iron-based alloy could be different from the diffusivity of nitrogen in pure iron. Presence of nitride precipitates could act as an obstacle for nitrogen diffusion in ferrite and therefore, diffusivity of nitrogen in ferrite of iron-based alloy could be smaller than diffusivity of nitrogen in pure iron [17].

There are two types of diffusion processes: (i) steady-state diffusion and (ii) non-steady state diffusion.

(i) *Steady-state diffusion*: In this case, concentration gradient is constant and diffusing flux does not change with time, for example, diffusion of a gas through a plate of metal (of thickness d) for which pressure (or concentration) of gas is held constant (C_1 and C_2) on both sides of the plate (Fig. 1.5). For the constant values of C_1 and C_2, if thickness of the plate is increased, diffusing flux will decrease. Flux of the diffusing species is given by Fick's first law:

$$J = -D\frac{dC}{dX} \tag{1.32}$$

Flux, J, is a measurement of the number of atoms per unit area that cross a particular plane per unit time. D is the diffusivity of the diffusing species.

(ii) *Non-steady state diffusion*: Most practical diffusion situations are non-steady state (e.g. nitriding, carburizing), i.e., concentration gradient, dC/dX, is a function of time, t (Fig. 1.6). Change in the concentration with depth and time is give by Fick's second law:

$$\frac{dC}{dt} = D\frac{\partial^2 C}{\partial X^2} \tag{1.33}$$

In Eq. (1.33), D is assumed to be independent of concentration. The solution to Eq. (1.33) for unidirectional diffusion from one medium to another across a common interface is of the general form:

$$C(X,t) = A - B\,\mathrm{erf}\left(\frac{X}{2\sqrt{Dt}}\right) \tag{1.34}$$

where A and B are constants to be determined from the initial and boundary conditions of a particular problem. Here, the diffusion direction X is perpendicular to the common interface between the surface-alloying atmosphere and specimen surface. The origin for X is at the interface. The two media are taken to be semi-infinite, that is, only one end of each of them (which is at the interface) is defined. The other two ends are at an infinite distance.

The solution of Eq. (1.33) for carburizing of steel can be obtained by considering the following bounder conditions (see also Fig. 1.6):

$$\begin{aligned}
\text{For } t = 0, &\quad C = C_0 \text{ at } x > 0 \\
\text{For } t > 0, &\quad C = C_s \text{ at } x = 0 \\
&\quad C = C_0 \text{ at } x = \infty
\end{aligned}$$

which gives

$$C(X,t) = C_s - (C_s - C_0)\,\mathrm{erf}\left(\frac{X}{2\sqrt{Dt}}\right) \tag{1.35}$$

Fig. 1.5 Fick's first law
demonstrated by the diffusion
of a gas through a plate of
metal with thickness d

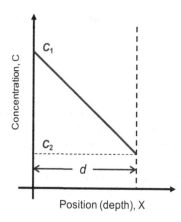

Fig. 1.6 Fick's second law
demonstrated by the change
in concentration-depth profile
with time

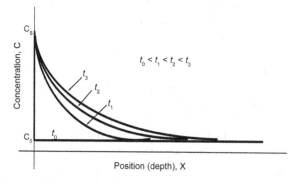

The first proposals to describe diffusion zone growth upon nitriding are based on a simple model originally meant for "internal oxidation" [18]. This model can be applied to internal nitriding by considering the following assumptions [19]:

(i)   The nitrogen dissolved in the ferrite matrix (α) exhibits Henrian behavior. This implies that the diffusion coefficient of nitrogen in the ferrite matrix is independent of the dissolved nitrogen content.

(ii) The reaction of dissolved nitrogen with dissolved alloying element (Me), leading to the nitride MeN_n, takes place only and completely at a sharp interface between the nitrided zone and the non-nitrided core.

(iii) The amount of nitrogen required for building up the concentration profile in the ferrite matrix of the nitrided zone is negligible in comparison to the amount of nitrogen consumed at the reaction interface.

(iv) Diffusion of Me can be neglected and is not nitriding-rate determining.

(v)   Local equilibrium prevails at the nitriding medium/specimen interface, so that the surface concentration $c_{N_\alpha}^S$ is equal to the lattice solubility of nitrogen, as given by the chemical potential of nitrogen in the nitriding atmosphere.

If these conditions are satisfied, it follows for the nitriding depth, X, at constant temperature:

$$X^2 = \left(\frac{2 \cdot c_{N_\alpha}^S \cdot D_N}{n \cdot c_{Me}} \right) \cdot t \tag{1.36}$$

where c_{Me} is the atomic concentration of substitutional solute Me originally dissolved, D_N is the diffusivity of nitrogen, and n is the atomic ratio N/Me in the nitride phase formed by nitriding.

This model has been often used to predict the case depth of the nitrided zone [17, 19–22]. However, a sharp interface between nitrided and unnitrided zone, as required by the model, does not occur in reality [19, 20]. Further, a major simplification of reality in the model is that the solubility of nitrogen in the ferrite matrix is taken as that pertaining to unstrained, pure α-Fe: the presence of excess nitrogen dissolved in the α-Fe matrix is not accounted for (see Sect. 1.8.1). These shortcomings of the simple model have been overcome in the numerical model proposed in Refs. [17, 20].

Henry's law holds for the nitrogen dissolved in α-Fe and hence it follows from Eq. (1.15) that $c_{N_\alpha}^S \sim a_N^S \sim r_N$. Then, considering the highly simplified model yielding Eq. (1.36), it can be suggested that the nitriding depth must be approximately proportional to $(r_n)^{1/2}$. However, at low nitriding potentials, nitriding of iron-based alloys could be lower than the expected [19]. At low nitriding potential, the value of $c_{N_\alpha}^S$ can be slightly lower than equilibrium solubility of nitrogen in the ferrite [5, 19]. This can be explained as follows. At relatively low-nitriding potential, a significantly large finite period of time is necessary to establish local equilibrium of the gas atmosphere with the solid substrate. This effect has been observed for nitriding of pure iron and is due to the finite rate of dissociation of NH_3 [23, 24]. The effect can be stronger for iron-based alloys at low-nitriding potential, because in the presence of Me (nitride forming alloying element) much more nitrogen has to be taken up before saturation at the surface can be attained. Apparently, for the iron-based alloys nitrided at low-nitriding potentials, lower thickness (than the expected) could develop if the nitriding time applied is too small to achieve saturation at the surface.

Case depth can be measured using various methods, like microstructural investigation of the cross-section of the surface alloyed specimens (using optical microscope and scanning electron microscope (SEM)), microhardness-depth profiling, and elemental concentration-depth profiling (using electron probe microanalysis (EPMA)), etc. [17, 19, 20]. If case depths are very thin, the specimen can be cut as shown in Fig. 1.7 with some angle, θ, and then such obtained cross-section will be investigated by using the methods mentioned above.

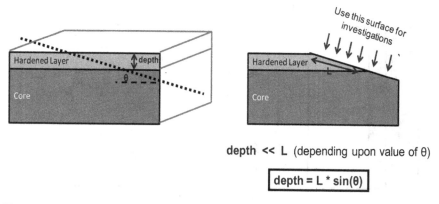

depth << L (depending upon value of θ)

$$depth = L * sin(θ)$$

Fig. 1.7 Scheme for determining case depth of thin layer

1.6 Nitrogen Concentration-Depth Profile of Nitrided Iron

To understand the phases present in the nitrided region and the corresponding nitrogen concentration-depth profile, it is necessary first to understand Fe–N phase diagram. Fe–N phase diagram contains the following phase fields: α (ferrite), γ (austenite), γ'-Fe$_4$N$_{1-x}$ and ε-Fe$_2$N$_{1-z}$. It is important to note that the Fe–N phase diagram shown in Fig. 1.8c [25] does not describe the equilibrium between Fe and N$_2$ at atmospheric pressure. This is because the solubility of nitrogen in iron is low, bond between N–N is strong in N$_2$ molecule (but, for successful nitriding of specimen, nascent/atomic nitrogen is required), and for equilibrium between Fe and N$_2$ requires very high pressure. The Fe–N phase diagram on this basis represents the equilibria between Fe and NH$_3$–H$_2$ mixture. Such equilibria can only occur at the interface of iron and the media concerned. Gaseous nitriding by using NH$_3$ is easier because at iron surface dissociation of NH$_3$ occurs easily to generate nascent nitrogen.

Layers of iron-nitrides can be formed at the surface of nitrided pure iron. Nitriding parameters play an important role in governing the formation of iron-nitrides during nitriding. Depending on the applied nitriding temperature and nitriding potential the hexagonal ε-Fe$_2$N$_{1-z}$ and/or the cubic γ'-Fe$_4$N$_{1-x}$ phases are formed. If the nitriding conditions allow the formation of the ε phase, it will be formed at the surface (Fig. 1.8a). The γ' phase can occur then as a second layer below the ε layer. Below the iron nitride layers always a zone of α-Fe with nitrogen dissolved in the octahedral interstitial sites of the body centered cubic Fe parent lattice is present (diffusion zone). Schematic illustration of a typical nitrogen concentration-depth profile of an iron-nitride compound layer is shown in Fig. 1.8b. Nitrogen contents at the interface between ε/nitriding-atmosphere, ε/γ' and γ'/ε are indicated in nitrogen concentration-depth profile and Fe–N phase

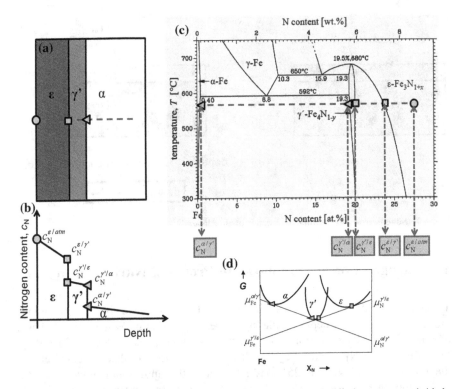

Fig. 1.8 a Schematic presentation of formation of iron-nitrides and diffusion-zone on nitrided iron (say, at about 580 °C). **b** Nitrogen concentration-depth profile along the *dashed-line* shown in (**a**). **c** Fe–N phase diagram [24] indicating nitrogen content, c_N), at 580 °C at the following interfaces: ε/nitriding-atmosphere, ε/γ' and γ/ε'. **d** Schematic presentation of free-energy *curves* at a constant nitriding temperature

diagram—where, $c_N^{\varepsilon/atm}$ is nitrogen content in ε in equilibrium with nitriding-atmosphere, $c_N^{\varepsilon/\gamma'}$ is nitrogen content in ε in equilibrium with γ', $c_N^{\gamma'/\varepsilon}$ is nitrogen content in γ' in equilibrium with ε, $c_N^{\gamma'/\alpha}$ is nitrogen content in γ' in equilibrium with α, and $c_N^{\alpha/\gamma'}$ is nitrogen content in α in equilibrium with γ'. At constant temperature, the amount of $c_N^{\varepsilon/atm}$ depends on the nitriding potential. At the ε/γ' (or γ'/α) interface, the chemical potential of nitrogen in both phases is equal (Fig. 1.8d), and there is an absence of two-phase region (e.g. $\varepsilon + \gamma'$ or $\gamma' + \alpha$) between the two single phase layers. Once iron-nitride formation occurs at the specimen surface during nitriding, nitrogen content in ferrite in equilibrium with γ' is independent of nitriding potential. Therefore, at constant nitriding temperature and time, thickness of the diffusion zone is independent of nitriding potential (see also Ref. [19]).

1.7   Improvement in Mechanical Properties Due to Surface Alloying

It is well known that the mechanical properties, like hardness, wear-resistance, and fatigue strength, of the workpiece are improved due to the surface alloying. These improvements in the properties are directly related the formation of new phases (e.g., compounds of alloying elements) and development of residual stresses/ strains in the surface layer during surface-alloying treatment. Residual stress/ strains are of two types: (i) macro-stress/strain and (ii) micro-stress/strain.

The development of macro-stress/strain during surface alloying is shown in Fig. 1.9. Consider the workpiece in contact with the nitriding or carburizing media. In the figure, N and C represent nitrogen and carbon, respectively. When the high chemical potential of N/C in the atmosphere is greater than the chemical potential of N/C in the workpiece, N/C will dissolve in the surface until equilibrium is established at the surface. Introduction of the external species, here N/C, into the surface causes the expansion of the surface layer. However, the expansion is resisted by the non-treated core. Therefore, compressive residual stress is developed in the surface and tensile stress in the immediate region of the core. The magnitude of the residual stress changes with change in the concentration of surface alloying element (Fig. 1.10). At a given process parameter of surface alloying treatment, surface has high concentration of element and it decreases with increase in the depth. Such concentration-depth profile leads to the maximum compressive stress at the surface and stress decreases with depth. Such behavior is observed in the low-temperature carburized austenitic stainless steel [26]. The presence of macro-stress/strain can be confirmed using X-ray measurements (Fig. 1.11). When X-rays with a particular wavelength (λ) interact with the polycrystalline crystalline specimen at a particular angle (θ), it diffracts in the specific direction by the lattice planes parallel to the specimen surface. A diffracted X-ray beam is determined by an interplanar spacing (d) at a specific lattice plane (hkl) in the specimen. In the x-ray diffraction method, by using characteristic x-rays with a unique wavelength λ, a diffracted angle θ can be obtained from a peak position of the intensity of X-ray diffraction. The well-known fundamental equation for X-ray diffraction phenomena for the first-order reflection from (hkl) plane is: $\lambda = 2d \sin (\theta)$. This is called Bragg's law. According to Bragg's law, the angle θ corresponds to the interplanar spacing d. Shift in the X-ray diffraction peak position in intensity versus 2θ plot occurs if macro-stress/strain is present at the surface (Fig. 1.11). Lattice strain ε at a lattice plane (hkl) is defined as the ratio of interplanar spacing d_0 at the non-strained state and d at the strained state: $\varepsilon = \frac{d-d_0}{d_0} = \frac{\sin(\theta_0)-\sin(\theta)}{\sin(\theta)}$. Figure 1.12 shows a specific (hkl) plane (say (111)) in various grains of surface region of polycrystalline material. When there is no residual macro-stress in the surface region, interplanar spacing ($d_{(hkl)}$) is the same in all grains. However, in the presence of compressive residual stress, interplanar spacing of (hkl) plane changes with the orientation of grain with respect to the

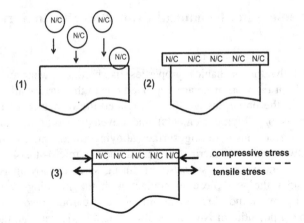

Fig. 1.9 Schematic presentation of the development of residual macro-stress/strain at the surface of workpiece during surface alloying (e.g., nitriding/carburizing). Here, N and C represent nitrogen and carbon respectively. Atoms of N/C are transferred from surrounding atmosphere to the surface of workpiece is shown in (*1*). (*2*) shows the free expansion of the surface layer without resistance from the core. But in a real situation, expansion of surface layer is resisted by the core which leads to the development of compressive residual stress in the surface layer while tensile residual stress in the core (*3*)

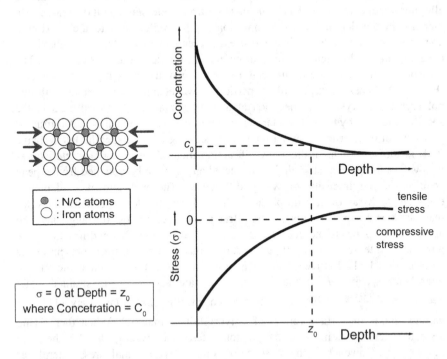

Fig. 1.10 Schematic presentation of the change in the residual macro-stress/strain with concentration of surface alloying elements

Fig. 1.11 Schematic X-ray diffraction peak recorded from the surface of non-treated specimen and surface alloyed specimen. Peak shifts leftward if compressive stress/strain is present and it shifts rightward if tensile stress is present

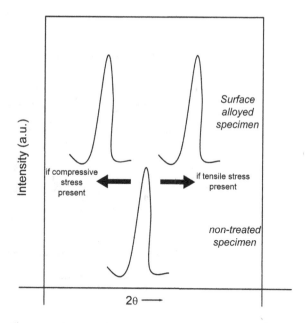

direction of compressive stress—interplanar spacing increases most for the planes parallel to the direction of stress and the spacing reduces most for the planes perpendicular to the direction of compressive stress. In order to confirm the residual macro-strain at the specimen surface, d_ψ versus $\sin^2\psi$ plot is widely used, where ψ is the angle of specimen tilt and d_ψ is the interplanar spacing of (hkl) plane when specimen is tilted by an angle of ψ (Fig. 1.13). As an example, X-ray diffraction peaks of $(311)_{\text{austenite}}$ recorded from the surface of carburized duplex stainless steel at various ψ angles is shown in Fig. 1.14. Increasing 2θ value with ψ, i.e. decreasing d_ψ with ψ, confirms the compressive macro-stress at the surface of carburized steel.

In many situations, surface alloying of ferrous alloys leads to the formation of new phases like nitride- or carbide- precipitates. If the lattice parameter of new phase is different from the matrix, misfit-strain field (micro-stress/strain) establishes surrounding the precipitates (Fig. 1.15a). The localized change in the lattice parameter of the matrix surrounding the precipitates leads to the broadening of X-ray diffraction peak of the matrix phase (Fig. 1.15b). The hardness of the metals can be defined as the resistance for plastic deformation. This resistance is directly related to the resistance for the motion of dislocations.[1] Presence of the hard precipitates in the soft matrix offers resistance to the motion of dislocation and

---

[1] Dislocation is one of the crystal defects in the material. If a plane ends abruptly inside a crystal we have a defect. However, the whole of abruptly ending plane is not a defect. Only the edge of the plane can be considered as a defect. This is a line defect. It is defined as the boundary between slip and no slip region of the slip-plane.

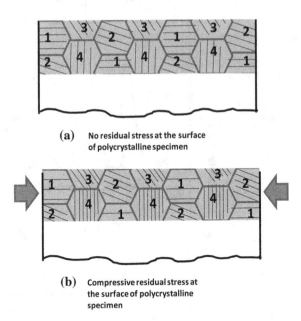

(a)   **No residual stress at the surface of polycrystalline specimen**

(b)   **Compressive residual stress at the surface of polycrystalline specimen**

Fig. 1.12 Surface region of polycrystalline specimen. A specific (hkl) plane is shown in various grains which are randomly oriented. **a** In the absence of any residual macro-stress, interplanar spacing of (hkl) plane in randomly oriented grains is constant (**a**). However, the compressive residual macro-stress changes the interplanar spacing of the same (hkl) plane (**b**). Grains indicated by number "1" and "4" have an orientation such that (hkl) plane is parallel and perpendicular, respectively, to the direction of compressive stress. In other grains indicated by numbers "2" and "3" (hkl) plane is inclined to the compressive stress direction

hence, hardness increases. The presence of strain field surrounding the precipitates further increases the resistance for the motion of dislocation. If coarsening of the precipitates occurs, coherent interface between precipitate and matrix is replaced by the incoherent interface. In such situation, strain field is relaxed and dislocations are formed along the interface.

Both macro- and micro-stress are responsible for the improvement in fatigue resistance because compressive stress at the surface delays the crack initiation while the latter helps in delaying the crack propagation. However, the improvement in hardness (and hence, wear resistance) is related to the formation of new hard phases (like precipitate) and micro-stress associated with them. If the precipitate and micro-stress formation does not occur, the improvement in wear resistance will not be significant; for example, shot-peening improves the fatigue resistance considerably, but the wear resistance is not affected much. Table 1.1 summarizes the properties and the cause for their improvements in surface alloyed workpieces.

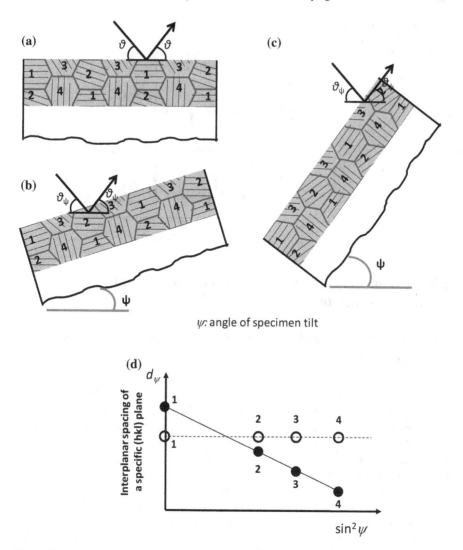

Fig. 1.13 Schematic presentation of the concept of $\sin^2\psi$ plot to determine residual macro-stress at the specimen surface (**a–c**). Interplanar spacing of (hkl) plane at various ψ angles is determined by X-ray diffraction peak position. **d** *Horizontal dashed-line* in d_ψ versus $\sin^2\psi$ plot indicates the absence of macro-stress at the specimen surface. However, negative slope of the solid-line in the plot is due to the compressive residual macro-stress at the specimen surface

Fig. 1.14 X-ray diffraction peaks of $(311)_{austenite}$ recorded from the surface of carburized duplex stainless steel at various ψ angles

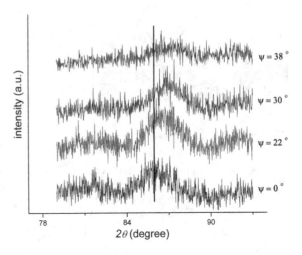

1.8 Some of the Interesting Researches in Nitriding and Carburizing

1.8.1 Nitriding of Fe–Me Alloys and Occurrence of Excess Nitrogen

During nitriding, if the iron matrix (substrate) contains alloying elements with a relatively high affinity for nitrogen, like Ti, V, Cr, Mn and Al, alloying element nitride precipitates can develop (in the "diffusion zone"), which leads to a pronounced increase of the hardness. The increase in, e.g., hardness or fatigue resistance depends on the chemical composition of the precipitates, their coherency with the matrix, their size, and their morphology.

If an Fe–Me (Me=Ti, V, Cr etc.) alloy is nitrided such that no iron nitrides can be formed at the surface (i.e., the nitriding potential is sufficiently low), only a diffusion zone containing MeN_n precipitates develops ("internal nitriding") [19, 20]. The nitrided zone is composed of alloying element nitride precipitates and surrounding α-Fe (ferrite) matrix containing nitrogen at octahedral interstitial sites. For MeN_n precipitate in α-Fe the nitride platelets have the orientation $(001)_{\alpha\text{-Fe}}//$ $(001)_{MeNn}$, which is compatible with the Bain orientation relationship (e.g. see Ref. [27] for VN).

Nitrided Fe-Me alloys have considerable capacity for the uptake of so-called excess nitrogen, i.e. more nitrogen than necessary for (i) precipitation of all alloying element as nitride, $[N]_{MeN_n}$, and (ii) equilibrium saturation of the ferrite matrix, $[N]_{\alpha}^{0}$. The total amount of excess nitrogen can be divided into two types: mobile and immobile excess nitrogen.

A significant part of the excess nitrogen in nitrided binary iron-based alloys is adsorbed at the nitride/matrix interfaces, $[N]_{interface}$. A MeN_n precipitate with

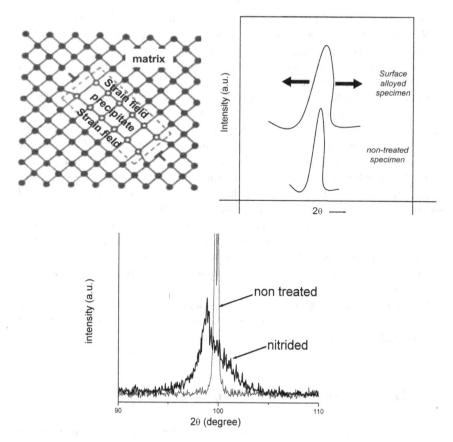

Fig. 1.15 a Schematic presentation of formation of precipitate in the matrix. Here, lattice parameter of the precipitate is different than the lattice parameter of matrix phase which leads to the formation of strain-field around precipitate. **b** Schematic X-ray diffraction peak recorded from the surface of non-treated specimen and surface alloyed specimen. Peak broadening occurs due to the presence of micro-stress/strain in the surface alloyed workpiece. **c** X-ray diffraction peak of $(211)_{\alpha\text{-Fe}}$ recorded from the surface of non-treated and nitrided Fe–V alloy

Table 1.1 Summary of the properties and the cause for their improvements in surface alloyed workpieces

Mechanical property	Reason for the improvement in property
Fatigue resistance	(Macro-stress) + (Micro-stress)
Hardness	(Micro-stress) + (Mechanical Property of new phases formed) + (Dislocation locking by interstitial atoms) + (In case of sever plastic deformation, interaction between dislocations)
Wear resistance	Hardness + Surface topography (if altered by surface alloying)

excess nitrogen adsorbed at the interface with the matrix can be regarded as an MeN_X compound, i.e., (X-n) nitrogen atoms per MeN_X molecule are bonded/adsorbed to the coherent faces of the particle/platelet. The amount of nitrogen at the precipitate/matrix interface is called immobile excess nitrogen as it does not take part in the kinetic process.

The coherent MeN_n precipitates induce strain fields in the surrounding ferrite matrix and thereby influence the nitrogen solubility of the ferrite matrix [28]. An MeN_n precipitate developing in the ferrite matrix experiences a positive volume misfit. Then, supposing fully elastic accommodation, the treatment by Eshelby [29] for a finite matrix shows that a positive dilation of the matrix occurs. The matrix lattice dilation generated by the misfitting nitrides, induced by the hydrostatic component of the image-stress field of finite bodies, provides a geometrical understanding for the occurrence of enhanced solubility of nitrogen. This dilation is not a direct function of temperature. The actually occurring, temperature-dependent amount of excess nitrogen (at strain fields: $[N]_{strain}$) can then be estimated applying the thermodynamics of (hydrostatically) stressed solids [28]. The amount of nitrogen at strain fields is called mobile excess nitrogen as it takes part in the diffusion (kinetic) process.

Total nitrogen uptake by nitrided Fe-Me system can be subdivided as shown in Fig. 1.16. Nitrogen absorption isotherms can be a useful tool to understand the various kinds of differently (chemically) bonded nitrogen.

1.8.2 Low-Temperature Carburizing

One of the interesting developments in surface alloying is the low-temperature carburizing developed by researchers in the USA [26, 30]. Highlights of some the results obtained by them are as follows.

Conventional gas carburization of stainless steels is performed at up to 1,010 °C. Hardness values can be increased from ~ 200 HV to between 700 and 750 HV. The improvement in hardness is due to the formation of chromium-carbides in the microstructures. However, the formation of chromium-carbides affects corrosion property because chromium is no longer available for the formation of passive chromium-oxide film. The precipitation of chromium carbides requires diffusion of substitutional elements (here, chromium). Substitutional element diffusion can be many orders of magnitude slower than interstitial diffusion of carbon. Therefore, a processing temperature window exists where significant carburization depths can be achieved while kinetically suppressing carbide formation. Low-temperature carburization of 316 stainless steel at 470 °C had produced carbon concentration more than 10 at.% while maintaining single-phase austenite. As a result of this, hardness of 1,200 HV, surface compressive residual stress values of ~ 2.1 GPa and enhanced corrosion resistance were observed.

Fig. 1.16 Flow chart indicating subdivision of nitrogen uptake in nitrided Fe-Me alloys. The amount of $[N]_{dislocation}$ can be neglected in recrystallized samples due to the relatively low dislocation densities

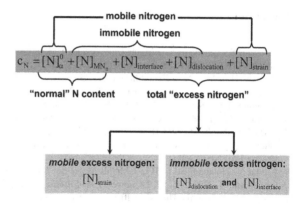

1.9 Conclusions

- The fatigue, tribological, and/or anti-corrosion properties of less expensive grades of alloys are possible to improve by using surface alloying treatments such as carburizing, nitriding, and chromizing. The mechanism of the surface alloying involves the following three important steps: absorption of diffusing species at specimen surface, inward diffusion of the absorbed species, and formation of compounds in the surface-treated region.

- Composition of the surface, formation of the phases, and case depth depend strongly on the temperature and chemical potential of the species (here, nitrogen, carbon and chromium) in atmosphere surrounding the specimen.

- Improvements in the mechanical properties of the surface due to the surface alloying are directly related to the formation of new phases (e.g., compounds of alloying elements) and the development of residual macro-/micro-stresses/ strains in the surface layer. Due to surface alloying, compressive residual macro-stress is developed in the surface and tensile stress in the immediate region of the core. The magnitude of the residual stress changes with change in the concentration of surface alloying elements. Both macro- and micro-stress are responsible for the improvement in fatigue resistance. However, the improvement in hardness (and hence, wear resistance) is related to the formation of new hard phases (like precipitates) and micro-stress associated with them. If the precipitates and micro-stress formation do not occur, the improvement in wear resistance will not be significant.

- A few of the interesting results about nitriding of binary Fe–Cr and Fe–V are the presence of "excess nitrogen" and two-types of nitride-precipitation morphologies in the nitrided region. In case of carburizing, the concept of low-temperature carburizing of stainless steels, developed by the researchers in the USA, is innovative. Such surface treatment on austenitic stainless steel resulted in the surface hardness of 1,200 HV, surface compressive residual stress values of ~2.1 GPa and enhanced corrosion resistance.

References

1. Hosmani SS (2013) Metall Mater Eng (former J Metalurgija) 19:65
2. Somers MAJ (2000) Heat Treat Met 27:92
3. Grabke HJ (1975) Arch Eisenhüttenw 46:75
4. Mittemeijer EJ, Slycke JT (1996) Surf Eng 12:152
5. Mittemeijer EJ, Somers MAJ (1997) Surf Eng 13:483
6. Budinski KG (1998) Surface engineering for wear resistance. Prentice hall, Englewood Cliffs
7. ASM Handbook 4 (1991) Heat treating. ASM International, Materials Park, OH
8. Ghosh A, Chatterjee A (2008) Ironmaking steelmaking: theory and practice. PHI Learning Private Ltd, New Delhi
9. Kumar V, Hosmani SS (2011) J Metall Mater Sci 53:393–404
10. Gressmann T, Nikolussi M, Leineweber A, Mittemeijer EJ (2006) Scripta Mater 55:723
11. Kooj BJ (1995). PhD Thesis, Delft University of Technology, The Netherlands
12. Porter DA, Easterling KE (1981) Phase transformations in Metals and alloys. Chapman & Hall, London
13. Wriedt HA, Gokcen NA, Nafziger RH (1987) Bulletin of Alloy Phase Diagrams 8:355
14. Grabke HJ (1968). Ber. Bunsengesell. Physic. Chem. 72 (in German):533
15. Callister WD (2007) Materials science and engineering: an introduction. John Wiley & Sons Inc, NY
16. Fast JD, Verrijp MB (1954) J Iron Steel Ins 176:24
17. Hosmani SS, Schacherl RE, Mittemeijer EJ (2007) Metall Mater Trans 38A:7–16
18. Meijering JL (1971) Advances in Material Research. Wiley Interscience, New York
19. Hosmani SS, Schacherl RE, Mittemeijer EJ (2005) Mater Sci Technol 21:113
20. Schacherl RE, Graat PCJ, Mittemeijer EJ (2004) Met & Mat Trans 35A:3387
21. Mittemeijer EJ, Vogels ABP, van der Schaaf PJ (1980) J Mater Sci 12:3129
22. Hekker PM, Rozendaal HCF, Mittemeijer EJ (1985) J Mater Sci 20:718
23. Rozendaal HCF, Mittemeijer EJ, Colijn PF, van der Schaaf PJ (1983) Metall Trans 14A:395
24. Friehling PB, Poulsen FW, Somers MAJ (2001) Z Metallkd 92:589
25. Liedtke D, Baudis U, Boßlet J, Huchel U, Klümper-Westkamp H, Lerche W, Spieß HJ (2006) Wärmebehandlung von Eisenwerkstoffen – Nitrieren und Nitrocarburieren. Expert-Verlag, Renningen Malmsheim
26. Michal GM, Ernst F, Kahn H, Cao Y, Oba F, Agarwal N, Heuer AH (2006) Acta Mater 54:1597
27. Bor TC, Kempen ATW, Tichelaar FD, Mittemeijer EJ, van der Giessen E (2002) Phil Mag 82A:971
28. Somers MAJ, Lankreijer RM, Mittemeijer EJ (1989) Phil Mag 59A:353
29. Eshelby JD (1956) Solid State Phy 3:79
30. Cao Y, Ernst F, Michal GM (2003) Acta Mater 51:4171

Chapter 2
Nitriding of Binary Iron-Based Alloys: An Overview

Abstract Nitrided iron-based alloys are widely used in various industrial applications. Composition of such alloys is one of the factors in governing their nitriding behavior. This chapter reviews some of the interesting results associated with nitrided binary iron-based alloys. It was observed that the concentration of alloying elements influenced the precipitation morphologies in nitrided zone of Fe–Cr and Fe–V alloys. Nitrided zone of such binary alloys showed the presence of "excess nitrogen" (i.e., actual nitrogen content was more than the normal/ expected nitrogen content). The occurrence of excess nitrogen was explained by using nitrogen absorption isotherms and kinetics of nitriding studies. The results obtained for the nitrided Fe-7 wt%Cr and Fe-4 wt%V alloys suggested that the compound layer, which appears "white" under optical microscope, was not just a pure γ'-Fe$_4$N, but was a combination of γ' plus nitrides of alloying element.

Keywords Fe–Cr alloy · Fe–V alloy · Nitriding · Discontinuous precipitation

2.1 Introduction

Nitriding of iron-based alloys is widely practiced in industries to enhance fatigue, wear, and anticorrosion properties of engineering components. The nitriding alloys used for these applications contain many alloying elements. Therefore, it is difficult to understand the nitriding behavior of the alloys caused by a particular alloying element. In this regard, a sequential study of nitriding of pure iron, binary alloys, etc., could be useful. Information about nitriding of pure iron is provided in detail in Refs. [1, 2]. A systematic understanding of the nitriding behavior of iron-based alloys has evolved over many years due to research by various authors, for example, see Refs. [3–5] for nitriding of Fe–Ti alloys, Refs. [6–12] for nitriding of Fe–Al alloys, Refs. [13–22] for nitriding of Fe–V alloys, Refs. [23–27] for nitriding of Fe–Cr alloys, and Refs. [28, 29] for nitriding of ternary alloys. Some of the interesting phenomena for nitriding of binary iron-based alloys, especially, Fe–Cr and Fe–V alloys, are summarized in this chapter.

S. S. Hosmani et al., *An Introduction to Surface Alloying of Metals*,
SpringerBriefs in Manufacturing and Surface Engineering,
DOI: 10.1007/978-81-322-1889-0_2, © The Author(s) 2014

2.2 Precipitation Morphologies in Nitrided Fe–Me Alloys

Nitriding of Fe-based alloys is generally carried out in the temperature range of 500–600 °C. During nitriding of the alloy, the surface of the specimen achieves equilibrium with the nitriding atmosphere and nitrogen dissolution in the ferrite (α-Fe) occurs. Dissolved nitrogen starts diffusing in the ferrite. Interaction of nitride forming alloying element with diffusing nitrogen leads to the formation of nitride precipitate. Precipitation of nitride in the ferrite matrix needs to overcome the activation energy barrier for nucleation. Morphology of the precipitate depends on the precipitate–matrix interface energy and lattice or volume misfit strain energy. For small misfit, the total change in free energy of the system is smaller for coherent/semi-coherent precipitates than incoherent precipitates. However, as precipitate size increases, incoherent precipitate could be associated with smaller free energy change of the system than coherent/semi-coherent precipitates [30]. In other words, the driving force for precipitate coarsening is the decrease in free energy of the system caused by the decrease in total precipitate–matrix interface area (for a constant volume fraction of the precipitates before and after coarsening) and misfit strain energy. Such coarsening is associated with decrease in the hardness of the alloy.

Nitrided Fe–Cr and Fe–V alloys showed two types of precipitation morphologies: (i) continuous precipitation and (ii) discontinuous precipitation.

These precipitations are defined as follows: (i) Continuous precipitation involves the transport of atoms to the growing nuclei of precipitating phase in the supersaturated matrix. This transport (diffusion) of atoms occurs over relatively large distances in the parent phase (i.e., long-range diffusional process). During this process, the average composition of the parent phase changes continuously toward its equilibrium concentration. (ii) Discontinuous precipitation involves structural and compositional changes in the region immediately adjacent to the advancing precipitate–matrix interface. In this precipitation, the composition of parent phase remains unchanged until it is consumed by the advancing precipitate–matrix interface.

Light optical micrographs of nitrided Fe-7 wt%Cr and Fe-4 wt%V alloys are shown in Figs. 2.1 and 2.2. Two types of the grains/regions are observed in the nitrided region—bright and dark. The bright region is adjacent to the non-nitrided core. However, the surface region is fully occupied by the dark grains. Micrographs are useful in understanding the possible sequence of events at a particular depth during the growth of nitrided zone. The sequence of events could be as follows: nontreated core → bright region → dark region. This suggests that the formation of bright region occurs initially which later transforms into dark region. Interestingly, the formation of dark region in the nitrided zone of Fe–Cr and Fe–V alloys depends on the concentration of alloying elements. Formation of dark region in the nitrided zone does not occur in the alloys containing up to 1 wt%Cr in case of Fe–Cr alloys (see Fig. 2.2) and 2 wt%V in case of Fe–V alloys (see Figs. 2.1 and 2.3). Nitriding for long duration does not transform bright region into

hardness indentation
within dark region

hardness indentation
within bright region

Fig. 2.1 Optical micrograph showing two hardness indentations (performed with the same load) in one original grain of the thick nitrided Fe-4 wt%V alloy: the *dark* region of the grain exhibits a larger indentation (i.e., possesses lower hardness) than the *bright* part of the grain. (Reprinted from Hosmani et al. [21], with permission from Elsevier)

dark region in the nitrided Fe-1 wt%Cr and Fe-2 wt%V alloys. These results suggest that high concentration of alloying elements promotes the formation of dark region in the nitrided zone. This is confirmed by using scanning electron microscopy (SEM), in Refs. [21, 24], where the dark region has lamellar morphology. However, optical microscopy and SEM are incapable to reveal any precipitation morphology in the bright region. X-ray diffraction studies [21, 24] have confirmed that the dark region consists of α-Fe and alloying-element-nitride phase (i.e., CrN and VN in nitrided Fe–Cr and Fe–V alloys, respectively). However, transmission electron microscopy studies [18, 31] have confirmed the presence of nitride precipitates in the bright region. Dark grains have considerably lower hardness than bright grains as demonstrated by the microhardness measurements (Figs. 2.1, 2.2, 2.3). On the basis of observed results, the bright region was designated as "continuous" precipitated region and dark region was designated as "discontinuous" coarsened region.

The mechanism for the discontinuous coarsening in nitrided Fe–V and Fe–Cr alloys is explained in Refs. [21, 24]. The essence of this mechanism is summarized below.

During nitriding, diffusing nitrogen interacts with vanadium (in Fe–V alloys) or chromium (in Fe–Cr alloys). This interaction initially leads to the formation of vanadium nitride or chromium nitride precipitates which are finely dispersed in α-Fe (ferrite) matrix ("continuous" precipitation). These precipitates are in the form of platelets (e.g., plate-like precipitates of VN are typically about 40 Å long with thickness of about 10 Å) which are probably coherent/semi-coherent with the α-Fe matrix [18, 31]. The lattice parameters of VN, CrN, and α-Fe are 0.4137, 0.4149, and 0.2866 nm, respectively. Due to the difference in lattice parameters of nitride and ferrite, the misfit-induced stress fields associated with the precipitates [18] are responsible for the high hardness values associated with "continuous" precipitated

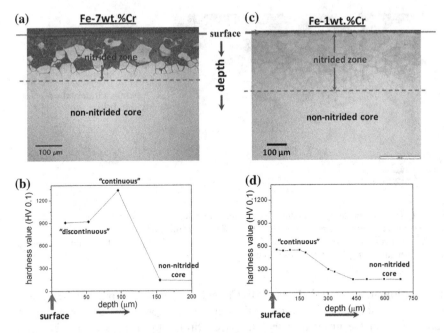

Fig. 2.2 Optical micrograph and corresponding microhardness-depth profile for **a** Fe-7 wt%Cr specimen nitrided at 580 °C for 4 h at a nitriding potential of 0.16 atm$^{-1/2}$ and **b** Fe-1 wt%Cr specimen nitrided at 580 °C for 5 h at a nitriding potential of 0.10 atm$^{-1/2}$

Fig. 2.3 **a** Light optical micrograph of the etched cross-section of the thick (thickness of 1.0 mm) Fe-2 wt%V specimen nitrided at 580 °C for 10 h at a nitriding potential of 0.10 atm$^{-1/2}$. **b** Hardness-depth profile of the same specimen. (Reprinted from Hosmani et al. [21], with permission from Elsevier)

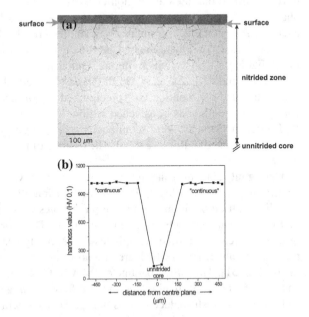

region (i.e., bright region of optical micrograph). A coarsening of the nitride precipitates occurs during prolonged nitriding treatment. The driving force for coarsening is the reduction in Gibbs free-energy caused by relaxation of internal long-range stress fields and reduction of precipitate–matrix interfacial area. This coarsening in general could be continuous growth of the precipitates ("Ostwald ripening"), but for the Fe–V and Fe–Cr alloys containing high concentration of alloying elements, coarsening occurs by a "discontinuous coarsening" of the former continuous precipitates. This discontinuous coarsening involves the growth of α-Fe and VN lamellae from nucleation sites like surfaces and grain boundaries. The following reaction possibly operates during discontinuous coarsening of nitride precipitates:

$$\alpha^| + \beta^| \rightarrow \alpha + \beta \qquad\qquad (2.1)$$

where $\beta^|$ denotes coherent nitride precipitates in a supersaturated (with nitrogen) ferrite matrix ($\alpha^|$). The reaction consists of replacing the submicroscopical and coherent/semi-coherent nitride precipitates by nitride lamellae ($\beta^| \rightarrow \beta$) with simultaneous elimination of the nitrogen supersaturation of the ferrite matrix ($\alpha^| \rightarrow \alpha$).

Therefore, on nitriding of Fe–V and Fe–Cr alloys that contain high concentrations of alloying elements, the following two reaction fronts can be expected: (i) initially, inward diffusion of nitrogen and the reaction of alloying element with N to form finely dispersed coherent/semi-coherent nitride precipitates and then (ii) subsequent discontinuous coarsening of the former continuous precipitates (see Figs. 2.1 and 2.2). As mentioned above, the hardness of showing discontinuous precipitated region is significantly lower than that of the continuous precipitated region. This behavior is a direct consequence of coarsening with associated loss of coherency. The larger the concentration of alloying element in the alloy, the larger the volume fraction of the nitride precipitates in the ferrite matrix. Therefore, the driving force for the discontinuous coarsening reaction will be larger for alloys containing higher concentration of alloying element. This concept explains the presence of dark region in nitrided Fe-4 wt%V alloy, but the absence of such region in nitrided Fe-2 wt%V alloy. Similar nitriding behavior in Fe–Cr alloys is observed.

It must be noted that the above-mentioned continuous and discontinuous precipitation morphologies are not reported in the literature for nitrided Fe–Ti [3–5] and Fe–Al [6–12] alloys.

2.3 Excess Nitrogen

The phenomenon of "excess nitrogen" in nitrided iron-based alloys was introduced in Chap. 1. The quantity of "excess nitrogen" is the difference between actual (experimental) nitrogen content and expected (theoretical) nitrogen content in the nitrided alloys. Expected nitrogen content in the nitrided Fe–Me (Me = Cr/V/Al/Ti) alloy is summation of the nitrogen necessary for precipitation of all alloying elements as nitride, $[N]_{MeN_n}$, and equilibrium saturation of nitrogen in the unstrained ferrite matrix, $[N]_\alpha^0$, at the given nitriding temperature and nitriding potential.

The total amount of nitrogen absorbed by the alloy is summation of various types of nitrogen. These types of nitrogen can be understood by using nitrogen absorption isotherm [7, 22, 32], which shows the dependence of the amount of nitrogen taken up by a (homogeneously) nitrided specimen as a function of the nitriding potential (directly related to the chemical potential of the nitriding atmosphere) [22]. Nitrogen uptake is determined by weight measurements of the specimens before and after the treatments. This method involves the following major steps:

(i) *Pre-nitriding*: this involves nitriding of the foil specimen such that complete thickness of the specimen is homogeneously nitrided.
(ii) *De-nitriding*: After pre-nitriding, the specimen is de-nitrided in pure H_2 at lower temperature than pre-nitriding temperatures. De-nitriding removes all types of nitrogen present in the specimen except the nitrogen which is strongly bonded to alloying element to form nitride phase.
(iii) *Determination of nitrogen absorption isotherms*: After the consecutive pre- and de-nitriding treatments, nitrogen absorption isotherms are determined at temperatures below the pre-nitriding temperature. Lower temperatures are selected to avoid any change in the precipitation morphology.

Nitrogen content in pre- and de-nitrided Fe-1.04 at.%Cr specimens are shown in Fig. 2.4. The difference between nitrogen content in pre-nitrided specimen and normal nitrogen uptake is the "total excess nitrogen." Figure 2.5 shows the nitrogen absorption isotherms for pre- and de-nitrided Fe-1.04 at.%Cr specimens at 560 °C and at 530 °C. Total nitrogen uptake in the nitrided specimen can be broadly classified into three types (I, II, and III) of nitrogen [7, 22, 32]:

(i) Type I: This is the strongly bonded nitrogen to alloying element in the corresponding stoichiometric nitride phase. This nitrogen cannot be removed easily by de-nitriding in a pure H_2 atmosphere. Type I nitrogen is indicated by level "A" in Fig. 2.5.
(ii) Type II: This is the adsorbed nitrogen at the nitride/matrix interface. This nitrogen is less strongly bonded compared to type I nitrogen. Therefore, this nitrogen can be removed by de-nitriding. Type II nitrogen is called *immobile excess nitrogen* because it does not take part in growth kinetics of the diffusion zone. This nitrogen corresponds to the difference between levels "B" and "A" in Fig. 2.5.
(iii) Type III: This is the dissolved nitrogen in the octahedral interstices of the ferrite matrix surrounding the precipitates. Type III nitrogen can be removed easily by de-nitriding. The amount of dissolved nitrogen in the ferrite matrix is directly proportional to the nitriding potential. Therefore, the straight-line dependence above level "B" in Fig. 2.5 represents the nitrogen dissolved

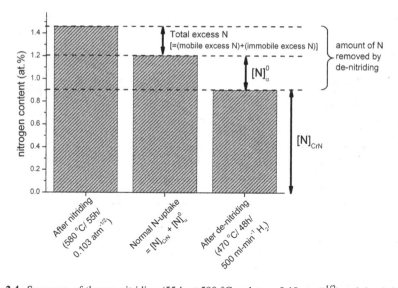

Fig. 2.4 Summary of the pre-nitriding (55 h at 580 °C and $r_n = 0.10$ atm$^{-1/2}$) and de-nitriding (48 h at 470 °C in pure H$_2$) experiments performed with Fe-1.04 at.%Cr foil specimens (thickness of 0.2 mm). [N]$_{CrN}$ is the amount of nitrogen required for the precipitation of all Cr as CrN and [N]$_\alpha^0$ is the amount of nitrogen dissolved in an unstrained ferrite matrix at the pre-nitriding conditions. (Reprinted from Hosmani et al. [32], with permission from Springer)

interstitially in the ferrite matrix. However, there is a difference between the nitrogen uptake in ferrite of nitrided alloy and unstrained pure iron. Higher nitrogen uptake in the ferrite of nitrided alloy is due to the presence of strain, which is caused by the misfit between nitride precipitate and ferrite matrix. [N]$_{strain}$ is also called *mobile excess nitrogen* because it diffuses to give excess thickness to the nitrided zone.

Interestingly, nitrogen content in the nitrided zone of Fe-4 wt%V alloy depends on the type of precipitation morphology [21]. Figure 2.6 shows the elemental concentration-depth profiles of the nitrided layer of the nitrided Fe-4 wt%V specimen. The presence of excess nitrogen in the nitrided zone is evident. Nitrogen content at the location of discontinuously transformed regions/grains is significantly lower than the continuously transformed regions/grains. In other words, the discontinuous transformation in nitrided Fe-4 wt%V alloy is associated with the loss in capacity of excess nitrogen. However, the nitrogen uptake in the nitrided zone of Fe–Cr alloys (containing 4, 7, 13, 20 wt%Cr) is independent of discontinuous or continuous precipitated regions [24–26]. The nitrogen-absorption isotherm for the discontinuously transformed Fe-20 wt%Cr alloy showed the presence of excess nitrogen in the nitrided specimen [27], i.e., the coarse and lamellar precipitation morphology of CrN/α-Fe is associated with considerable quantity of excess nitrogen. This excess nitrogen could be ascribed to an unexpected, minor fraction of the total Cr content in the alloy present as coherent, tiny nitride platelets within the ferrite

Fig. 2.5 Nitrogen-absorption isotherms as observed for pre- and de-nitrided Fe-1.04 at.%Cr specimens **a** at 560 °C and **b** at 530 °C. The nitrogen level after de-nitriding is indicated by "A." The linear portion of the absorption isotherm is indicated by the *solid line* that intersects the ordinate at $r_n = 0$ at a nitrogen level indicated by "B." Level B plus the nitrogen solubility in unstrained ferrite is shown by the *dotted line*. The lattice solubilities of nitrogen in pure iron, $[N]_\alpha^0$, as function of the nitriding potential at 560 and 530 °C were taken from Ref. [22]. (Reprinted from Hosmani et al. [32], with permission from Springer)

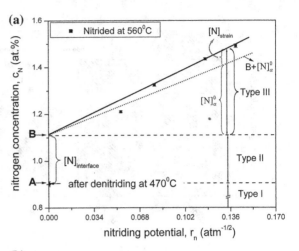

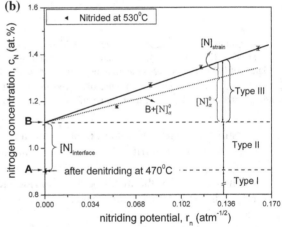

lamellae of the "discontinuously coarsened" lamellar precipitation morphology, as evidenced by transmission electron microscopy [27].

Nitriding kinetics study [21, 25] is also useful to understand the occurrence of excess nitrogen in nitrided alloys. By using the simple model originally proposed for "internal oxidation" (see Chap. 1), expected nitriding depth can be calculated. Experimental values for nitriding depth and total excess nitrogen can be obtained from electron probe microanalyzer (EPMA). Incorporating experimental nitriding depth in the model gives actual nitrogen content in the ferrite matrix at specimen surface. Difference between such obtained nitrogen content and the nitrogen in unstrained ferrite (calculated from Refs. [1, 2]) gives the *mobile excess nitrogen* content. Subtracting mobile excess nitrogen from total excess nitrogen gives *immobile excess nitrogen* content. Such calculations are summarized in Table 2.1 for nitrided Fe-4 wt%V and Fe-2 wt%V alloys [21]. In this way, the total amount of excess nitrogen can be divided into mobile and immobile excess nitrogen.

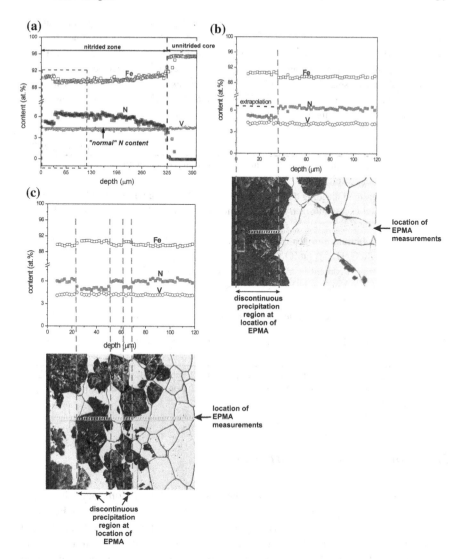

Fig. 2.6 a Elemental (N, V, Fe) concentration-depth profiles of Fe-4 wt%V alloy nitrided at 580 °C for 10 h at a nitriding potential of 0.10 atm$^{-1/2}$. The horizontal *gray line* indicates the "normal" amount of nitrogen uptake. **b** Enlargement of part of the depth profile, indicated by the *dashed rectangle* in (**a**), together with the light optical micrograph of the cross-section analyzed (which was etched after EPMA analysis) showing the location of the EPMA measurements (as revealed by the usual track of carbon contamination). Clearly, the nitrogen concentration in the surface adjacent *dark* part of the nitrided zone is smaller than in the *bright* part of the nitrided zone. **c** Part of elemental concentration-depth profile recorded at another location of the same specimen, together with the light optical micrograph of the cross-section analyzed (which was etched after EPMA analysis) showing the location of the EPMA measurements. Clearly, the nitrogen concentration at the location of *dark* grains ("discontinuously" transformed grains) is smaller than in the *bright* grains ("continuously" precipitated grains). (Reprinted from Hosmani et al. [21], with permission from Elsevier)

Table 2.1 Tabular presentation of the expected nitriding depth calculated using "internal oxidation" model (see Chap. 1) with $c_{N_\alpha}^S = c_{N_\alpha}^{S,0}$, the measured nitriding depth determined from the experimental (EPMA) data, the total amount of excess nitrogen at the surface determined from EPMA, the equilibrium nitrogen content in (stress-free) α-Fe ($c_{N_\alpha}^{S,0}$), the nitrogen content in α-Fe at the surface calculated using "internal oxidation" model with experimental values for the nitriding depths ($c_{N_\alpha}^S$), the amount of dissolved, mobile excess nitrogen ($= c_{N_\alpha}^S - c_{N_\alpha}^{S,0}$), and the amount of immobile (adsorbed at precipitate /matrix interfaces) excess nitrogen

		Fe-4 wt%V	Fe-2 wt%V
(1)	Expected nitriding depth ("internal oxidation" model with $c_{N_\alpha}^S = c_{N_\alpha}^{S,0}$) ($\mu$m)	210	296
(2)	Measured nitriding depth (experimental (EPMA) results) (μm)	322	425
(3)	Total excess nitrogen at surface (measured from EPMA) (at.%)	2.31[a]	1.15
(4)	Equilibrium nitrogen content in α-Fe, $c_{N_\alpha}^{S,0}$ (calculated from Refs. [1, 2]) (at.%)	0.24	0.24
(5)	Nitrogen content in α-Fe at the surface, $c_{N_\alpha}^S$ ("internal oxidation" model with experimental values for nitriding depths) (at.%)	0.56	0.49
(6) = (5) − (4)	Dissolved (i.e., mobile) excess nitrogen ($= c_{N_\alpha}^S - c_{N_\alpha}^{S,0}$) (at.%)	0.32	0.25
(7) = (3) − (6)	Adsorbed precipitate/matrix interfacial (i.e., immobile) excess nitrogen (at.%)	1.99	0.91

All results have been given for the Fe-4 wt% V and Fe-2 wt% V alloys, nitrided at 580 °C for 10 h at a nitriding potential of 0.10 atm$^{-1/2}$. (Reprinted from Hosmani et al. [21], with permission from Elsevier)

[a] Estimate; see Ref. [21]

2.4 Compound-Layer Formation

In nitriding, compound-layer is also known as whiter-layer. Compound-layer can be made up of γ'-Fe$_4$N and/or ε-Fe$_3$N. Compound-layer formation over Fe-based alloys depends on the chosen nitriding parameters, like temperature and nitriding potential. At a given nitriding temperature, the formation of compound-layer can be controlled by selecting appropriate nitriding parameters. Effect of nitriding potential on nitriding kinetics of Fe-7 wt%Cr was investigated at constant temperature (580 °C) and time (4 h) [26]. The Fe-7 wt%Cr specimens nitrided at low nitriding potentials ($r_n = 0.01$–0.16 atm$^{-1/2}$) do not show compound-layer formation. However, at high nitriding potentials ($r_n = 0.21$–0.82 atm$^{-1/2}$), γ'-Fe$_4$N layer formation occurs. The nitriding potential has great influence on the depth of the nitrided zone. The nitriding depth depends approximately linearly on the square root of the nitriding potential for the specimens nitrided in the range of nitriding potentials of 0.01–0.16 atm$^{-1/2}$. As soon as a closed iron-nitride layer forms on the surface, the dependence of the nitriding kinetics on the nitriding potential becomes marginal. At constant temperature and time, growth kinetics of

Fig. 2.7 Summary of the important aspects of the nitriding behavior of Fe–Cr and Fe–V alloys

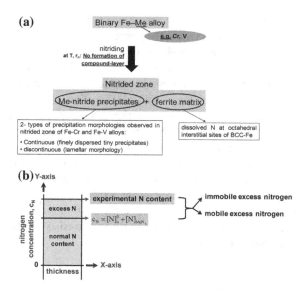

the diffusion zone is dependent on the solubility of nitrogen in ferrite matrix at the specimen surface. Once there is a formation of compound-layer on the specimen surface, the amount of nitrogen dissolved in the ferrite matrix at the interface between compound-layer and diffusion zone becomes independent of nitriding potential. The growth of compound-layer with nitriding potential is insignificant. Therefore, the overall nitriding depth (diffusion zone plus compound-layer thickness) becomes independent of nitriding potential once the compound-layer forms on the specimen's Fe-7 wt%Cr surface.

Nonuniform thickness and irregular penetration of the compound-layer in the diffusion zone are observed for nitrided Fe-7 wt%Cr and Fe-4 wt%V alloys [33, 34]. Phase analysis (by using x-ray diffraction) and quantitative analysis (by using electron probe microanalysis) of the compound-layer demonstrated the presence of alloying element nitride phase within γ'-Fe$_4$N matrix [33, 34]. These findings for nitrided Fe–Cr and Fe–V alloys suggest that the compound-layer, which appears "white" under optical microscope, is not just a pure γ', but is a combination of γ' plus nitrides of alloying element.

2.5 Conclusions

- Important aspects of the nitriding behavior of Fe–Cr and Fe–V alloys are summarized in the form of a flowchart in Fig. 2.7.
- Nitriding of Fe–Cr and Fe–V alloys showed two types of nitride precipitation morphologies in the diffusion zone: continuous and discontinuous. The discontinuously precipitated region appears dark under optical microscope and has

lamellar morphology. Discontinuous precipitation depends on the concentration of alloying elements. This precipitation morphology is observed for nitrided alloys containing high concentration of alloying elements (more than about 2 wt% of alloying elements). However, these precipitation morphologies are not reported in the literature for nitrided Fe–Ti and Fe–Al alloys.

- Nitrided Fe–Me (Me = Cr/V/Al/Ti) alloys show the presence of excess nitrogen in the nitrided zone. Excess nitrogen can be well understood by using nitrogen absorption isotherms. Nitriding kinetics study is also useful to understand the occurrence of excess nitrogen in nitrided specimens. It has been realized that the total amount of excess nitrogen is the combination of two types of excess nitrogen: mobile and immobile. Mobile excess nitrogen is responsible for greater nitriding depth than the expected value.
- White-layer formed on the surface of nitrided Fe-7 wt%Cr and Fe-4 wt%V alloys has nonuniform thickness and consists of alloying element nitride phase within γ'-Fe$_4$N matrix.

References

1. Mittemeijer EJ, Slycke JT (1996) Chemical potentials and activities of nitrogen and carbon imposed by gaseous nitriding and carburizing atmospheres. Surf Eng 12:152–162
2. Mittemeijer EJ, Somers MAJ (1997) Thermodynamics, kinetics, and process control of nitriding. Surf Eng 13:483–497
3. Jack DH (1976) Acta Metall 24:137
4. Podgurski HH, Davis FN (1981) Acta Metall 29:1
5. Rickerby DS, Henderson S, Hendry A, Jack KH (1986) Acta Metall 34:1687
6. Podgurski HH, Oriani RA, Davis NA (1969) Trans Metall Soc AIME 245:1603 (with Appendix by Li JCM, Chou YT)
7. Biglari MH, Brakman CM, Mittemeijer EJ, van der Zwaag S (1995) Phil Mag 72A:931
8. Biglari MH, Brakman CM, Somers MAJ, Sloof WG, Mittemeijer EJ (1993) Z Metallkd 84:124
9. Steenaert JS, Biglari MH, Brakman CM, Mittemeijer EJ, van der Zwaag S (1995) Z Metallkd 86:700
10. Biglari MH, Brakman CM, Mittemeijer EJ, van der Zwaag S (1995) Metall Mater Trans 26A:765
11. Biglari MH, Brakman CM, Mittemeijer EJ (1995) Phil Mag 72A:1281
12. Meka S, Hosmani SS, Clauss AR, Mittemeijer EJ (2008) Int J Mater Res 99:808
13. Philipps A, Seybolt AU (1968) Trans Metall Soc AIME 242:2415
14. Pope M, Grieveson P, Jack KH (1973) Scand J Metall 2:29
15. Welch WD, Carpenter SH (1973) Acta Metall 21:1169
16. Krawitz A (1977) Scr Metall 11:117
17. Yang MM, Krawitz AD (1984) Metall Trans 15A:1545
18. Bor TC, Kempen ATW, Tichelaar FD, Mittemeijer EJ, van der Giessen E (2002) Phil Mag 82A:971
19. Djeghlal ME, Barrallier L (2003) Ann Chim Sci Mat 28:43
20. Gouné M, Belmonte T, Redjaimia A, Weisbecker P, Fiorani JM, Michel H (2003) Mater Sci Eng 351A:23
21. Hosmani SS, Schacherl RE, Mittemeijer EJ (2005) Acta Mater 53:2069
22. Hosmani SS, Schacherl RE, Mittemeijer EJ (2006) Acta Mater 54:2783

23. Hekker PM, Rozendaal HCF, Mittemeijer EJ (1985) J Mater Sci 20:718
24. Schacherl RE, Graat PCJ, Mittemeijer EJ (2002) Z Metalld 93:468
25. Schacherl RE, Graat PCJ, Mittemeijer EJ (2004) Metall Mater Trans A 35:3387
26. Hosmani SS, Schacherl RE, Mittemeijer EJ (2005) Mater Sci Technol 21:113
27. Hosmani SS, Schacherl RE, Lityńska-Dobrzyńska L, Mittemeijer EJ (2008) Phil Mag 88:2411
28. Clauss AR, Bischoff E, Hosmani SS, Schacherl RE, Mittemeijer EJ (2009) Metall Mater Trans 40A:1923
29. Jung KS, Schacherl RE, Bischoff E, Mittemeijer EJ (2010) Surf Coat Technol 204:1942
30. Porter DA, Easterling KE (1981) Phase transformations in Metals and alloys. Chapman & Hall, London
31. Vives Díaz NE, Hosmani SS, Schacherl RE, Mittemeijer EJ (2008) Acta Mater 56:4137–4149
32. Hosmani SS, Schacherl RE, Mittemeijer EJ (2008) J Mater Sci 43:2618–2624
33. Hosmani SS, Schacherl RE, Mittemeijer EJ (2006) Int J Mater Res (formerly Zeitschrift für Metallkunde) 97:1545–1549
34. Hosmani SS, Schacherl RE, Mittemeijer EJ (2009) J Mater Sci 44:520–527

Chapter 3
Influence of Process Parameters in Plasma-Nitriding, Gas-Nitriding, and Nitro-Carburizing on Microstructure and Properties of 4330V Steel

Abstract The effect of different operating parameters of plasma-nitriding, gas-nitriding, and nitro-carburizing of 4330V steel (NiCrMoV low alloy high strength steel) was studied. Fluidized-bed furnace was used for gas-nitriding and nitro-carburizing. Cold wall pulse dc unit was used for plasma-nitriding. The specimens were gas-nitrided/nitro-carburized at 500–580 °C for 4 or 6 h, and plasma-nitrided at 460–550 °C for 5–28 h. Surface treated specimens were characterized using optical microscope, scanning electron microscope, X-ray diffraction, and microhardness measurement. Results showed that uniformity of "white-layer" depends significantly on the type of processing. The white-layer was more uniform in plasma-nitriding compared to gas-nitriding and nitro-carburizing. Under the influence of temperature, the white-layer thickness increased for all three types of treatment. However, the thickness was lowest for plasma-nitrided specimens. Thickness of the white-layer was in the range of 2–11 μm for plasma-nitrided, 25–36 μm for gas-nitrided, and 29–40 μm for nitro-carburized specimens. All surface-treated specimens showed the formation of iron-nitride networks along the grain boundaries at sharp edges and corners. Along with increase in temperature, the penetration of white-layer into the diffusion zone increased. In gas-nitrided specimens, some areas showed penetration of iron-nitride up to depth of 246 μm. Surface hardness of the gas-nitrided and nitro-carburized specimens was 690 and 750 $HV_{0.1}$, respectively. However, plasma-nitrided specimens showed hardness values in the range of 550–620 $HV_{0.1}$. In case of the nitro-carburized specimen, white-layer was composed of only ε-Fe_3N phase, which suggests that the presence of carbon in the furnace atmosphere helped in stabilizing ε-Fe_3N rather than γ'-Fe_4N. In case of the plasma-nitrided specimens, temperature and gas ratio had influence on the constituents of the white-layer.

Keywords Nitriding · Nitro-carburizing · Crankshaft · Iron-nitride network

S. S. Hosmani et al., *An Introduction to Surface Alloying of Metals*,
SpringerBriefs in Manufacturing and Surface Engineering,
DOI: 10.1007/978-81-322-1889-0_3, © The Author(s) 2014

3.1 Introduction

Plasma-nitriding or iron-nitriding is an advanced surface processing technology that has experienced substantial industrial development over the past 30 years. An area where fine nitriding control is required, plasma-nitriding provides a better solution because it allows close control over the composition and case depth of nitride structures. A relatively low operating temperature (400–540 °C) prevents dimensional distortion and hence, eliminating the post machining work.

Plasma-nitriding is different from other processes in that no direct heat input is given to the component to be nitrided and consequently, the temperature rise is attributed to the plasma formation. A major advantage of plasma-nitriding lies in control of the brittle and hard compound-layer (ε-Fe$_3$N and γ'-Fe$_4$N) thickness (typically, it is less than 10 μm) and therefore extensively improves properties like wear, fatigue, corrosion, and load-bearing capacity of dynamically loaded components. It is observed that components undergoing cyclic loads, viz., crankshafts are prone to develop cracks at the surface after certain number of cycles and these cracks provide failure initiation sites. The main reason for formation of these cracks has been stated as the brittle compound-layer at the surface [1].

Plasma-nitriding makes use of the glow discharge technology in which the component to be nitrided is kept at a negative potential with respect to furnace wall (anode) and a potential difference is built between them keeping the furnace chamber at low pressure of the order of 10^{-4} to 10^{-3} mbar. These operating conditions ionize the H$_2$ gas thereby heating the component due to bombardment of H$^+$ ions and intermolecular collisions between them; in the later stage, nitrogen is introduced and high temperature permits ionization of N$_2$ into N$^+$ ions and these ions diffuse into the specimen. Typical gas mixtures used are N$_2$:H$_2$::1:4 to ensure optimum nitriding potential ($r_N = P_{N2}/P_{H2}$) so that the compound-layer thickness can be controlled and at the same time the required case depth is achieved. Plasma-nitriding can be divided into two methods depending on the ability to control the current and voltage [1, 2]:

(i) *Continuous DC* This method uses a fixed value of input current depending on the surface area of the specimen and the voltage changes with respect to pressure. The maximum heat generation is achieved through kinetic energy created due to bombardment of the gas atoms. In this method, temperature increase is directly proportional to voltage up to a certain range; once the desired temperature is attained voltage is kept constant and consequently the temperature remains constant. The disadvantage of this method lies in the inability to switch off the power supply if any arcing occurs.

(ii) *Pulsed DC* Critical component geometry needs the ability of the process to switch on and off the power supply as required to ensure uniform charge buildup. The DC peak voltage can be varied according to the part geometry based on the term called "duty cycle". In this method, pulse frequency can be adjusted, and due to this, the power supply can be turned off or on between the frequency limits and hence, temperature uniformity can be achieved.

Nitro-carburizing is one of the processes of surface alloying where carburizing atmosphere is introduced in the nitriding atmosphere. The advantages of the process are as follows [2]: (i) in addition to nitrogen, carbon is also introduced into the surface of specimen, (ii) apart from the dissociation of NH_3, an additional source for nitrogen is hydrogen cyanide (HCN), which is generated due to the interaction between nitriding and carburizing gases, (iii) it is possible to obtain mono-phase white-layer over the specimen surface, and (iv) kinetics of the process improves significantly compared to gas-nitriding.

3.1.1 Nitriding and Nitro-carburizing of 4330V Steel: Motivation and Objectives

Nitriding has been used extensively to improve the hardness, fatigue, and wear-resistant properties for steels over decades. Although a lot of research has been carried out on gas-nitriding, nitro-carburizing, and the newly developed plasma-nitriding, some problems still remain unanswered. In spite of a close control over the operating conditions and nitriding the specimen in controlled conditions, uniformity in white-layer thickness over the entire specimen has not been achieved [1, 2].

Along with this, the problem of nonuniform surface hardness is experienced in many components like gears, crankshafts, etc. In most of the cases, "white-layer" penetrates deeper into the diffusion zone, which is not acceptable, because in case of fatigue these areas can propagate further and eventually lead to premature fracture of the component [3].

Geometry of the component also plays an important role in nitriding; at sharp corners and edges, saturation of nitrides takes place and this forms a network. This problem was first seen in case of gears. At corners where this network was formed, chipped off during loading, and hence disturbing the gear tooth profile. The effect of different types of treatment on the same steel is also important because from this data, specific nitriding cycle for a given steel and component can be designed [4, 5].

Gas composition also affects the thickness of white-layer and the hardness values of the nitrided specimen hence its effect on the properties of nitrided layers has to be studied. Depending on the gas composition, the composition of white-layer is decided and ultimately the properties that will be achieved [4].

White-layer is a hard and brittle phase and many applications demand for situations where no white-layer is desired; in these cases, the white-layer has to be removed after nitriding by mechanical ways and hence this adds to the work time. If the nitriding potential is controlled, nitriding can be done in the absence of white-layer where only diffusion zone is present [6].

The current work focused on surface alloying of 4330V steel, which is an NiCrMoV low alloy high-strength steel. This steel is mainly used for components subjected to dynamic loading, for example, crankshafts used in oil industry components. The objectives of the current work are as follows:

1. To study the effect of specimen geometry on nitriding behavior of steel.
2. To study the white-layer formation on nitrided/nitro-carburized steel surface.
3. To study the effect of temperature kinetics of the white-layer formation.
4. To study the effect of gas composition in gas-/plasma-nitriding/nitro-carburizing atmosphere on microstructure.

3.2 Experimental

3.2.1 Specimen Preparation

Alloy used in this project was 4330V, which was a NiCrMoV low alloy high-strength steel. The chemistry of this alloy is given in Table 3.1.

The crankshafts made of the 4330V alloy had undergone the following heat treatment cycle, prior to nitriding:

- *Stress relieving* Soaking at 600 °C for 8 h, followed by furnace cool to 300 °C and finally, air cooling.
- *Hardening* Holding at 845 °C for 5 h and then, followed by oil quenching.
- *Double tempering* (i) Holding at 630 °C for 10 h, followed by furnace cooling to 426 °C and finally, air cooling; (ii) Holding at 580 °C for 8 h and then, air cooling.

The surface-alloyed specimens were cut with the help of abrasive cut machines. These specimens were then embedded (with the help of cold setting compound and mold powder) such that specimen cross-section was parallel to the embedded mold surface. Mold preparation was necessary for easy handling of small specimens, which were cut from the surface-alloyed specimens. The cross-section of the specimens was prepared by polishing (last step: 3 μm alumina slurry) and etching with 3 % nital.

3.2.2 Surface Alloying

3.2.2.1 Plasma-Nitriding

Initial trials were conducted on the actual crankshafts. These trials were conducted at different temperatures, time, and varying gas ratios. Along with the crankshafts, test coupons (see Fig. 3.1) were also loaded on fixtures.

Cold-wall plasma-nitriding furnace vessel and all parts to be loaded in the furnace were cleaned by acetone. The crankshaft and test coupon were loaded on a

Table 3.1 Chemical composition of 4330V alloy

C (wt%)	Si (wt%)	Mn (wt%)	P (wt%)	S (wt%)	Cr (wt%)	Mo (wt%)	Ni (wt%)	V (wt%)
0.32 ± 0.02	0.25 ± 0.10	0.88 ± 0.13	0.03	0.01	0.85 ± 0.11	0.45 ± 0.05	1.83 ± 0.18	0.08 ± 0.03

Fig. 3.1 Geometry of the test coupon used in the plasma-nitriding experiments

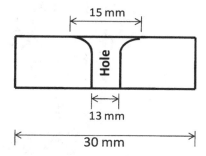

horizontal plate, which was connected to negative potential. After loading, the furnace's upper door was lowered (Fig. 3.2) and an insulating wax was applied on the joint to avoid air gaps that can otherwise disturb the vacuum. Initially, a rotary pump had drawn all the air from the vessel to achieve dynamic vacuum. The lowering of the pressure was recorded on a display panel and accordingly, the pump speed was maintained. Pressure was maintained at 2 mbar. In the next step, hydrogen gas was introduced into the furnace through an independent line and this flow was controlled by a Baratron isolation valve. Figure 3.2 shows the loading pattern of the crankshaft in the plasma-nitriding furnace.

Hydrogen was useful in attaining the required operating temperature. The typical mechanism of this heating is as follows. When hydrogen was introduced into the system, it ionized into H^+ ions and started bombarding on the crankshaft, which was at negative potential. Typically, the voltage inside the furnace was between 840 and 900 V and current was between 30 and 40 A. At such high values of current and voltage, bombarding of the hydrogen ions heated the workpiece and the surrounding atmosphere. As there were no external heating elements, the time required to attain the nitriding temperature, e.g., 520 °C, was about 6–8 h.

After attainment of the temperature, nitrogen gas was introduced into the system through another independent line so that it ionized into N^+ ions and subsequently, diffusion started. At the end of nitriding cycle, flow of nitrogen gas was stopped and hydrogen was gradually decreased. Once the temperature decreased to 300 °C, the pressure inside the chamber was lowered and slowly brought back to atmospheric pressure.

Some of the plasma-nitriding experiments were performed on the test coupons (Fig. 3.1), without the presence of crankshaft. For such smaller loads inside the chamber, it was difficult to achieve pressure of 2 mbar and successful nitriding (possible reason could be as follows: when the rotary pump drew air from the vessel, such smaller parts could fall from the cathodic plate and, moreover, sufficient glow discharge could not be established, which caused unsuccessful

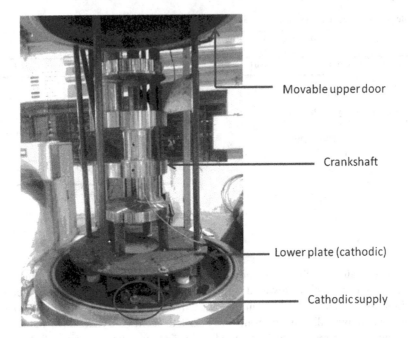

Movable upper door

Crankshaft

Lower plate (cathodic)

Cathodic supply

Fig. 3.2 Loading pattern of crankshaft in the cold-walled plasma-nitriding furnace

nitriding). In this situation, series of steel bars called dummy loads were placed into the furnace to compensate for the low weight. This arrangement was successful to accomplish the nitriding cycle.

In this work, nine plasma-nitriding experiments were performed. Specimen designation and corresponding nitriding parameters are summarized in Table 3.2.

3.2.2.2 Gas-Nitriding and Nitro-Carburizing

Gas-nitriding and nitro-carburizing were carried out in fluidized-bed furnace (Fig. 3.3a). Fluidized-bed furnace consists of a fluidization chamber containing a mixture of sand and alumina particles. Ammonia, air, nitrogen, and LPG were the gases required for nitriding or nitro-carburizing in the furnace. The primary use of nitrogen in fluidized-bed furnace was to obtain the optimum fluidization of sand and alumina particles so that uniform temperature could be maintained surrounding the specimens.

While operating the furnace, initially the air valve was opened and air was introduced into the chamber by turning on the air compressor. The required temperature was set by the temperature controller and the flow rate of air was adjusted such that optimum fluidization was achieved. Flow rate was recorded on the air column. As the set temperature was increased, the optimum flow rate was decreased. Generally, in the case of fluidized-bed furnace, as the temperature

Table 3.2 Specimen designation and process parameters for the experiments performed in the plasma-nitriding furnace

Designation	Gas ratio (N₂:H₂)	Temperature (°C)	Pressure (mbar)	Voltage (V)	Current (A)	Duty cycle (%)	Time (h)
Job No. 1	1:4	526	1.2	840–855	32	10	26
Job No. 2	1:3	535	1.2	890–870	28	7	28
Job No. 3	1:3	545	1.2	870–840	28	7	28
Job No. 4	1:3	560	1.2	840–830	34	7	34
Job No. 5	1:4	530	1.2	840–830	34	7	30
Job No. 6	1:4	530	1.2	805–790	28	7	20
Plasma 1	1:3	500	1.6–1.8	1040–1050	50	8	8
Plasma 2	1:1	460	2.6	690	14	8	5
Plasma 3	1:3	510	2.2	940	14	8	5

(a)

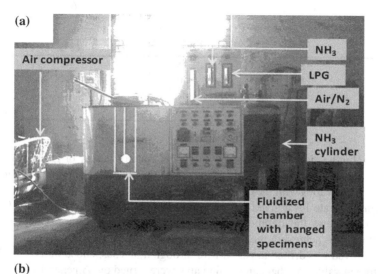

(b)

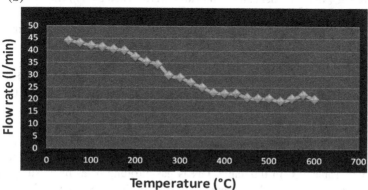

Fig. 3.3 a Fluidized-bed nitriding/nitro-carburizing furnace. **b** The flow rate of air needed for optimum fluidization at various temperatures

Table 3.3 Specimen designation and process parameters for the experiments performed in fluidized-bed nitriding/nitro-carburizing furnace

Job designation	Temperature (°C)	Gas ratio (NH_3:N_2)	Flow rate (l/min)		Time (h)
			N_2	NH_3	
GN1	520	1:3	19	6	4
GN2	540	1:3	19	6	6
GN3	550	1:3	19	6	6
GN4	560	1:3	19	6	6
GN5	580	1:3	19	6	6

Job designation	Temperature (°C)	Gas ratio (NH_3:Air)	Flow rate (l/min)		Time (h)
			Air	NH_3	
AN1	550	1:8	24	3	6
AN2	550	1:8	24	3	6
AN3	550	1:3	19	6	6

Job designation	Temperature (°C)	Gas ratio (%)			Time (h)
		N_2	NH_3	LPG	
NC1	530	60	37.5	2.5	4
NC2	540	50	45.5	4.5	6
NC3	550	60	34.5	5.5	6
NC4	560	60	33.5	6.5	6

increases, fluidization tends to increase and flow rate of the air or gases has to be decreased so that the sand does not come out of the chamber. Therefore, optimum flow rate of air varies with the desired nitriding temperature. Let us assume that this optimum flow rate for a given nitriding temperature is x liters per minute (l/min). Figure 3.3b shows the optimum flow rate versus set temperature. Once the set temperature was achieved, the air compressor was turned off. The specimen to be surface alloyed was drilled to pass a wire through it and suspended in the chamber. The nitrogen valve and then, ammonia valve were turned on. A general thumb rule applied here was as follows: the optimum air flow rate for the given nitriding temperature (x l/min) was equal to the sum of the nitrogen flow rate (x_1 l/min) and ammonia flow rate (x_2 l/min), i.e., $x = x_1 + x_2$ (where, $x_1 > x_2$). In case of nitro-carburizing, LPG flow rate (x_3 l/min) was incorporated, i.e., $x = x_1 + x_2 + x_3$.

A pilot lamp was fired on top of the fluidization chamber so that the gases coming out of the chamber were burned immediately. At the end of the heat treatment cycle, ammonia and LPG supply were closed, but nitrogen was allowed to flow to avoid oxidation of specimen. The heater was then turned off.

In this work, 12 gas-nitriding and nitro-carburizing experiments were performed. Specimen designation and corresponding process parameters are summarized in Table 3.3. In this table, GN is gas-nitriding of as-received specimen (which was hardened and tempered before nitriding: see Sect. 3.2.1), AN is nitriding of annealed specimen (annealing was at 850 °C for 6 h), and NC is nitro-carburizing of as-received specimen.

3.2.3 Specimen Characterization

3.2.3.1 Optical Microscopy

Microstructure of polished and etched samples were analyzed on Zeiss Axiovert 40MAT microscope and the micrographs were processed using Axio Vision software V12.1. Optical micrographs of the cross section of specimens were taken after microhardness measurements to calculate the perpendicular distance of hardness indentations from the surface. Optical micrographs were also used to measure the white-layer (compound-layer) thickness.

3.2.3.2 Hardness Measurements

Microhardness of the nitrided specimens was taken with a load of 100 g and 15 s dwell time. Along with surface, hardness readings were taken on the cross-section of the specimen until the core hardness was reached. This profile was used to determine the case depth. To minimize the possible error in the hardness values, precaution was taken that no two hardness indentations were closer than two times the diagonal of the indentations. Hardness readings on the crankshaft were taken with Equotip hardness tester or the Leeb Hardness tester.

3.2.3.3 Scanning Electron Microscope (SEM) and X-Ray Diffraction (XRD)

Scanning electron microscopy of the cross-section of surface-alloyed specimens was done using JSM 6360 Jeol Scanning electron microscope with accelerating voltage of 20 kV.

For phase analysis, X-ray diffraction patterns were recorded from the surface of surface-alloyed specimens using Braker AXS with Cu radiation ($\lambda = 1.5405$ Å). The indexing of the acquired XRD patterns was done using PCPDF and Xpert software.

3.3 Results and Discussion

3.3.1 Effect of Specimen Geometry

Initial experiments were performed to understand the effect of specimen geometry on plasma-nitriding behavior of crankshaft and the test coupon. For this purpose, hardness measurements were done at various locations on crankshaft and test coupons. Hardness readings were taken on journals and pins crankshaft. Three

Fig. 3.4 J1, J3, J5, and J7 are journals and P2, P4, and P6 are pins of crankshaft. Hardness measurements were performed on these journal and pins to confirm the uniformity of plasma-nitriding (see Table 3.4)

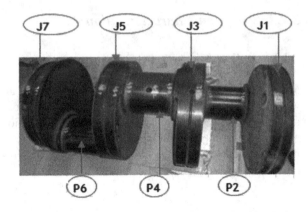

Table 3.4 Hardness at different locations on the surface of Job No. 1 plasma-nitride at 526 °C for 26 h by using the gas ratio of 1:4::N_2:H_2

Job No. 1				
Locations	Hardness in HR_C (Equotip readings)			
	A	B	C	Average
J7	47	45	47	46.3
J7 (near to collar)	38	36.8	40	37.2
P6 near to J7	35	40	38.4	37.8
P6 near to J5	39	37	37.6	37.86
J5	48	48	47.8	48
P4 near to J3	36	39	36.2	37
P4 near to J5	35	40	35.4	37.5
J3	48	48	49.4	48
P2 near to J3	37.7	38	35	36.9
P2 near to J1	48	39.2	35	40.7
J1	49.9	49.4	50.4	50

readings were taken on the circumference of each journal and pin at locations A, B, and C. These locations were separated by 120°. Figure 3.4 shows the locations of the hardness measurements on crankshaft. The hardness readings are summarized in Table 3.4. Considerable variation in the hardness values (from about 35 to 49 HRc) was observed on journals and pins.

Lowest hardness readings have been observed on locations, P2 (near to J1 and J3), P4 (near J3 and J5), and P6 (near J5 and J7). The locations with low hardness are the pins while the journals show considerably high hardness values. Crankshaft being a bi-axial component pins are the areas where cross-sectional area changes abruptly. Table 3.4 suggests that hardness readings as low as 35 HRc have been observed while the hardness of the material before nitriding ranges between 34 and 38 HRc. This confirms that the areas near the pins have not been nitrided.

In plasma-nitriding, it is reasonable to assume that the current density keeps on changing and therefore, affecting the glow envelope that surrounds the workpiece [1]. Surfaces with complex geometry dissipate heat at different rates, which implies that complex geometries are subjected to different kinetics when nitrided simultaneously. Nonuniformity of hardness on plasma-nitrided surface of complex geometries can be explained by the hollow cathode effect [1], which states that when two rods are placed close to each other in a plasma-nitriding furnace, the current density in the region between the two components increases and arc discharge takes place [2] leading to no nitriding in the region. In the case of crankshaft, the hollow cathode effect is probably the main cause behind low hardness values on the pin because pin is between two adjacent journals (for example, P6 is between J7 and J5: see Fig. 3.4), which behave like two different walls and the area between them gets overheated and the sputtering rate dominates the diffusion rates and hence, no nitriding takes place. This further suggests that the glow envelope is unable to follow the profile of the crankshaft.

As diffusion is a time and temperature-dependent phenomenon, temperature, time, and gas ratio were increased (Table 3.2) so that the pins (along with journals) could be nitrided. However, the issue of hardness variations (as mentioned above for Job No. 1) persists for Job Nos. 2, 3, and 4. According to the literature, the geometry of the workpiece could also affect the thickness and uniformity of the nitride layers [7].

To eliminate the issue of nonhomogeneity of surface hardness of plasma-nitrided crankshafts, a simple experiment (see further) was useful. Through this experiment, the glow could be concentrated in a specific region rather than compassing the whole furnace. In this experiment, a plate on top of the crankshaft and two plates parallel to the pins were mounted. The cathodic supply was extended from the base plate to the plate kept over the crankshaft so that the glow could be compressed between the bottom and the top plates. Figure 3.5 shows the modified setup. The next experiment was conducted in Job No. 5 using the modified setup. Hardness readings on this nitrided crankshaft are shown in Table 3.5. The obtained results confirm that the modified setup is useful to eliminate the surface hardness variation at various locations on pins and journals. Successive trials were taken with the same setup (with the same process parameters) to verify the reproducibility of the results, which also proved successful.

As stated earlier, test coupons were also plasma-nitrided along with the crankshaft. The geometry of the component plays an important role in plasma-nitriding and it influences the depth up to which nitrogen can diffuse and consequently the hardness values. In one of the plasma-nitrided experiments, the coupon was nitrided along with Job No. 1. After nitriding, the disk-shaped coupon was cut into two halves. As shown in Fig. 3.6a, the squared section was cut from the specimen and embedded in the mold material and hardness profiling was done along the directions as indicated by profiles 1 and 2. Microhardness-depth profiles at the two locations of the cross-section are shown in Fig. 3.6b.

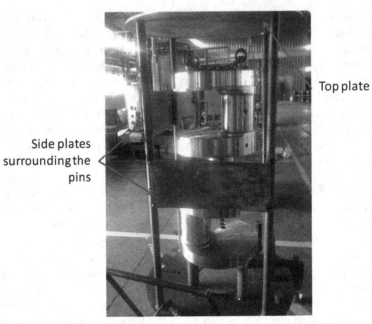

Top plate

Side plates
surrounding the
pins

Fig. 3.5 Modified plasma-nitriding setup for crankshaft with additional tope and side plates

Table 3.5 Hardness at different locations on the surface of Job No. 5 plasma-nitride at 530 °C for 30 h by using the gas ratio of $1:4::N_2:H_2$

Job No. 5				
Locations	Hardness in HR_C (Equotip readings)			
	A	B	C	Average
J7	49.4	47	47.1	47.8
J7 (near to collar)	47.8	50	48	48.6
P6 near to J7	46	48	47	47
P6 near to J5	48.3	47.2	47.2	47.5
J5	49	49	49.9	49.3
P4 near to J3	44	44	47.7	45.2
P4 near to J5	47.8	47.8	48	47.8
J3	49.6	49	50.9	49.3
P2 near to J3	47.8	48.9	49.5	48.7
P2 near to J1	50.2	49.5	49.7	49.8
J1	50.8	50.4	51.5	50.9

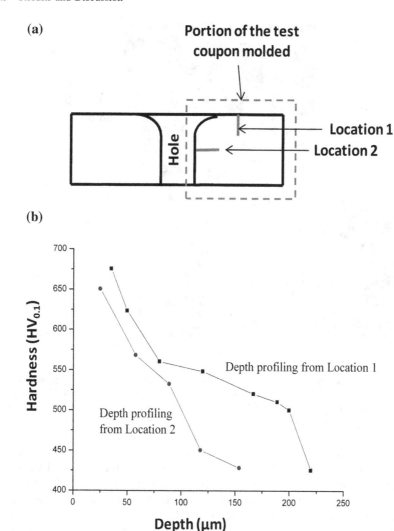

Fig. 3.6 **a** Vertical section of the plasma-nitrided (at 526 °C for 26 h by using the gas ratio of 1:4::N_2:H_2) test coupon indicating the locations of hardness-depth profile measurements. **b** Hardness-depth profiles at the locations indicated in (**a**)

3.3.2 Nonuniform Iron-Nitride Growth

3.3.2.1 Iron-Nitride Penetration in the Diffusion Zone

Although the hardness values and the thickness of iron-nitride layer (also, known as white-layer or compound-layer) are in compliance with the requirements during applications, problems encountered frequently for the nitride components are the

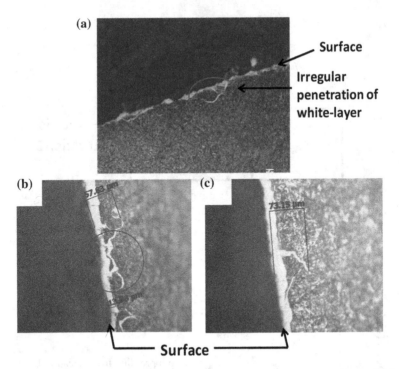

Fig. 3.7 Optical micrographs of the cross-section of the 4330V specimen **a** plasma-nitrided at 535 °C for 28 h by using the gas ratio of 1:3::N_2:H_2 (*specimen* test coupon with Job No. 2), **b** gas-nitrided at 560 °C for 6 h (*specimen* GN4), and **c** nitro-carburized at 550 °C for 6 h (*specimen* NC3). White-layer penetration is indicated in the micrographs

development of irregular white-layer and deep penetration of iron-nitride in the diffusion zone, especially at the interface region between white-layer and diffusion zone. Penetration of iron-nitride is generally observed along the grain boundaries. Since iron-nitride is brittle in nature, it makes the grain boundaries weaker and therefore, the white-layer can chip off during dynamic loading or prove as crack initiation sites. The micrographs of the cross-section of nitrided and nitro-carburized specimens are shown in Fig. 3.7. Irregular penetration of iron-nitride is present in plasma-nitrided, gas-nitrided, and in nitro-carburized specimens. It is observed that penetration of iron-nitride cannot be avoided by selecting surface-alloying method (plasma-, gas-nitriding, and nitro-carburizing). However, the density of these cracks and the length up to which they diffuse in the diffusion zone is very less in plasma-nitrided specimen compared to gas-nitrided and nitro-carburized specimens.

On nitriding, nitrogen diffuses through the ferrite grains by bulk diffusion which causes the formation of MN, where M is the alloying element. Irregular penetration of white-layer possibly suggests that, during the growth of white-layer, the interface between white-layer and diffusion zone is supersaturated with nitrogen (excess nitrogen) and iron-nitride starts nucleating along grain boundaries in the diffusion zone (due to the easy nucleation sites). The transformation process from ferrite to iron-nitride possibly starts at the grain boundaries and progresses toward the center of ferrite grains. Deep penetration of the white-layer could be due to (i) the dominating grain boundary diffusion over the bulk diffusion, (ii) the high misfit-stress in the diffusion zone near to the white-layer, which causes micro-cracks at the interface region and provides the path for nitrogen penetration, or (iii) the pre-existence of microcracks in the surface region during pre-nitriding heat treatment or machining process.

3.3.2.2 Iron-Nitride Network Formation at the Corner/Edge

Similar to the penetration of iron-nitride deep into the diffusion zone formed at the surface, iron-nitride network formation at corner/edge of the specimen is another problem observed in the nitrided and nitro-carburized 4330V steel (Fig. 3.8). There are evidences in the literature about the formation of iron-nitride network at the corner [1, 2]. The possible explanation and the effect of this network formation at corner are as follows [1, 2]: At the corner, nitrogen diffuses from all angles and causes the normal reaction of nitrogen saturation at the corner. If the nitriding process is allowed to proceed unchecked, the supersaturated solution of nitrogen can form at the corners, leading to nitride networking throughout the corner region. This network forms along the grain boundaries (i.e., iron-nitride envelops the grain boundaries at the corner). The net result is that the corner can become very brittle and prematurely fail by chipping or spalling during application of the nitrided component.

Figure 3.8 indicates the nonuniformity in the white-layer at corners and the iron nitrides form a cobweb-like structure. When components are subjected to loading, the circled areas are prone to cracking and later become crack initiation sites; this crack growth leads to chipping off of the corners. Hence, failure of the component takes place much below the fatigue limit.

In case of gas-nitriding, to understand the influence of nitriding potential and temperature on network formation at the corner, an experiment was performed at 520 °C for 6 h using the gas ratio of $1:5::NH_3:N_2$ (NH_3: 2 l/min and N_2: 20 l/min). In this experiment, gas ratio was intentionally kept low (i.e., the nitriding atmo-sphere was diluted—low nitriding potential) and also, temperature was reduced to ensure minimum possible diffusion. The effect of these operating parameters on the structure of nitrided layers is shown in Fig. 3.8b. It can be inferred from Fig. 3.8b that, although the thickness of white-layer is minimum, network for-mation is not avoided. However, by controlled nitrogen potential and/or temper-ature, minimum possible thickness of white-layer can be achieved and this can assist in minimizing the iron-nitride network at corners.

In case of plasma-nitriding, current and voltage play an important role in the formation of glow discharge (which is the nascent nitrogen carrying medium). When sharp corners are present in the specimen/component, they have a tendency to draw more current because different shapes dissipate different amounts of heat

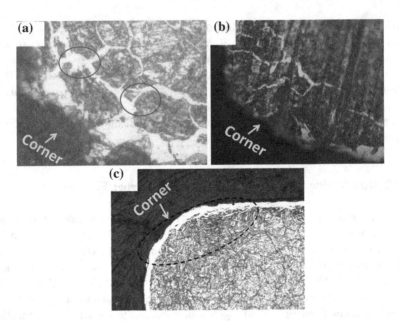

Fig. 3.8 Optical micrographs of the cross-section of the 4330V specimens **a** nitro-carburized at 530 °C for 4 h (*specimen* NC1), **b** gas-nitrided at 550 °C for 6 h (specimen: GN3), and **c** plasma-nitrided at 545 °C for 28 h by using gas ratio of 1:4::N$_2$:H$_2$ (*specimen* test coupon with Job No. 3). Iron-nitride network formation at the corner is indicated in the micrographs

[1]. The power input at these sharp edges is more compared to flat surfaces and hence at these points the temperature is slightly higher than that of the surroundings [1]. Due to this temperature gradient, more and more current is drawn toward the corners and therefore, it is possible that the diffusion rate of nitrogen at the corners is more and saturation of nitrogen can take place easily. Figure 3.8c shows the nitride network formed in plasma-nitride specimen.

3.3.3 Effect of Temperature

As nitriding is a diffusion process, temperature is the most important parameter in governing the depth at which the diffusing species can reach and thereby, control the case depth. In case of nitriding, after nitride precipitation of strong nitride-forming alloying elements, the only element left for nitrogen to combine with is Fe (of course, depending on the chosen nitriding potential). Iron being the major constituent, nitrogen readily combines with iron to form Fe$_3$N or Fe$_4$N, which is termed as white-layer or compound-layer. It is reasonable to assume that the higher the temperature, the greater will be the thickness of white-layer and consequently, the case depth.

3.3.3.1 Effect on White-Layer Thickness

Irrespective of the method of surface-alloying, white-layer thickness increases with increase in temperature. Figure 3.9 shows the optical micrographs of the etched cross-sections of the plasma-nitrided, gas-nitrided, and nitro-carburized 4330V specimens. Specimen plasma-nitrided at 526 °C (Fig. 3.9a) has white-layer thickness of about 4.75 μm and surface hardness of about 580 $HV_{0.1}$. Thicknesses of the white-layer for specimens plasma-nitrided at various temperatures are shown in Table 3.6. With increase in temperature from 460 to 560 °C, as expected, the thickness of white-layer increases from 1–2 μm to about 8 μm. Interestingly, it is observed that at higher temperatures, the white-layer (compound-layer) starts acquiring two phases, γ'-Fe_4N and ε-Fe_3N (see Sect. 3.3.4). In practice, a single phase compound-layer could be preferred because the γ' phase and ε phase have different crystal structures (FCC and HCP respectively) and if they are present in conjunction, the boundary between these two is strained and could easily be delaminated under small loading [1, 2].

As the plasma-nitriding temperature increases, the surface hardness increases. At process temperatures where the ε phase is dominant, specimen shows pronounced increase in the hardness (because ε phase is a brittle phase).

The white-layer thickness of gas-nitrided and nitro-carburized specimens is comparatively higher than the plasma-nitrided specimens at a particular temperature (Fig. 3.9a vs. Fig. 3.9b and for 560 °C, Table 3.6 vs. Fig. 3.9d). The possible reason behind this is explained as follows. In plasma-nitriding, diffusion of nitrogen in specimen takes place only when ionization occurs. These ionized species bombard the surface (sputtering action) and therefore, actual available species for nitriding are low. But, in nitriding and nitro-carburizing, nitriding species are generated due to the dissociation of ammonia and the available species for surface alloying are more.

SEM micrograph of the specimen (NC2) nitro-carburized at 540 °C is showed in Fig. 3.10. If compound-layer is composed of two phases (γ' and ε), different topographic contrast and sharp boundary between two phases could be observed in SEM micrograph layers (for example, see Ref. [8]). However, Fig. 3.10 shows the single phase compound-layer for nitro-carburized specimen. Presence of carbon in the nitro-carburizing atmosphere possibly enhances the contribution of ε phase in the compound-layer and the effect could be similar to the role of carbon content in steel in compound-layer formation during nitriding [1]. Surface hardness of the nitro-carburized specimen is about 750 $HV_{0.1}$.

3.3.3.2 Effect on White-Layer Penetration and Networking

The main effect of increasing temperature is on the extent of white-layer penetration in the diffusion zone and the intensity of iron-nitride network formation at corners/edges. General observation showed that the extent of white-layer

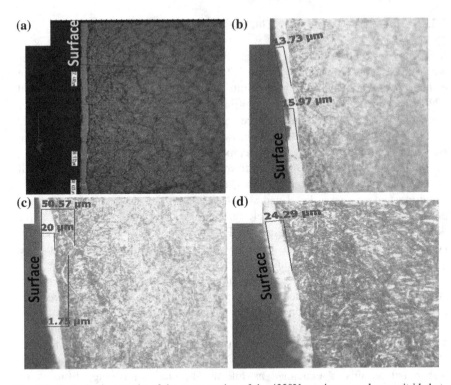

Fig. 3.9 Optical micrographs of the cross-section of the 4330V specimens **a** plasma-nitrided at 526 °C for 26 h by using the gas ratio of 1:4::N$_2$:H$_2$ (specimen: test coupon with Job No. 1), **b** gas-nitrided at 520 °C for 4 h (*specimen* GN1), **c** gas-nitrided at 580 °C for 6 h (*specimen* GN5), and **d** nitro-carburized at 560 °C for 6 h (*specimen* NC4)

Table 3.6 Thickness of white-layer of 4330V specimens plasma-nitrided at various temperatures

Specimen designation	Temperature (°C)	Time (h)	Average value of white-layer thickness (μm)	Average value of surface hardness (HV$_{0.1}$)
Plasma 2	460	5	2.1	not measured
Plasma 1	500	8	2.7	not measured
Plasma 3	510	5	3.5	not measured
Job No 2	535	28	4.6	600
Job No 3	545	28	6.5	614
Job No 4	560	34	8.9	620

penetration increases with increase in temperature. In Fig. 3.9b–c, optical micrographs of the specimens gas-nitrided at 520 and 580 °C illustrate the effect of nitriding temperature on irregularities in white-layer growth. For some specimens gas-nitrided at 560 °C, penetration of the iron-nitride was as deep as 244 μm in the

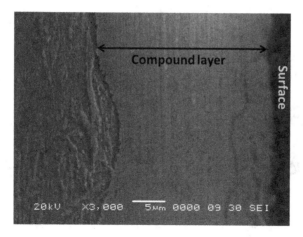

Fig. 3.10 Scanning electron micrograph (SEM) of the cross-section of specimen nitro-carburized at 540 °C for 6 h (*specimen* NC2). The micrograph shows mono-phase compound-layer

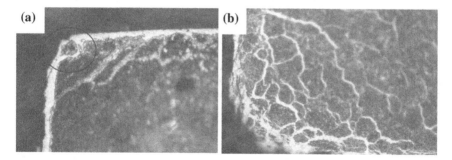

Fig. 3.11 Optical micrographs of the cross-section of the 4330V specimens plasma-nitrided **a** at 510 °C for 5 h by using the gas ratio of 1:3::N_2:H_2 (*specimen* Plasma 3) and **b** at 560 °C for 34 h by using the gas ratio of 1:3::N_2:H_2 (*specimen* test coupon with Job No. 4)

diffusion zone. In this regard, it could be beneficial to nitride at lowest possible temperatures depending on the case-depth requirement. Similar issue of iron-nitride penetration is also observed in the specimens nitro-carburized at high temperatures. However, in case of plasma-nitrided specimens, no such change in penetration depth of iron-nitride is seen and it remained more or less the same over a wide range of temperatures.

With increase in temperature and time, the intensity of iron-nitride network at corners increases (Fig. 3.11). It is observed that the small area of the corner is covered by nitride network at 510 °C/5 h, but the density of network increases considerably at 560 °C/34 h.

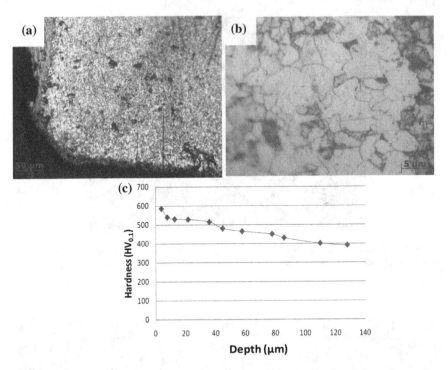

Fig. 3.12 Optical micrographs of **a** the cross-section at lower magnification, **b** the surface region at higher magnification, and **c** microhardness-depth profile of the annealed specimen nitrided at 550 °C for 6 h by using a gas ratio of 1:8::NH_3:Air

3.3.4 Effect of Gas Composition

Difference in the gas composition used during nitriding has influence on the phase constituents of nitrided layer and on the nitriding kinetics. Different gas compositions lead to different concentration of nitriding species in the nitriding atmosphere, i.e., gas composition of the nitriding atmosphere governs the nitriding potential (r_N). To study this effect, various experiments were performed on 4330V steel using different gas compositions.

Figure 3.12a–b shows the micrographs and Fig. 3.12c shows microhardness-depth profile of the annealed specimen nitrided at 550 °C for 6 h using gas ratio of 1:8:: NH_3:Air. In this experiment, the nitriding media is dilute with respect to the ammonia content and the nitriding potential is the lowest among all other gas-nitriding experiments carried out in this work. No white-layer formation is observed for this specimen (see also X-ray diffraction results below). But, surface hardness is about 580 $HV_{0.1}$ and hardness-depth profile confirms the presence of diffusion zone. It has to be noted here that air was used in the experiment instead of nitrogen gas and this causes the surface oxidation.

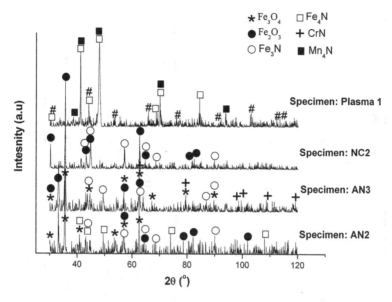

Fig. 3.13 X-ray diffractograms recorded from the surface of specimens nitrided with different gas compositions

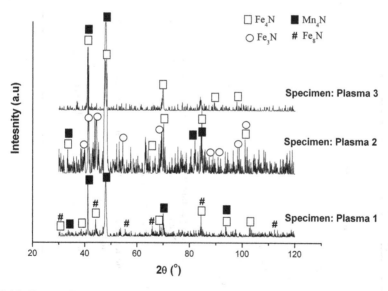

Fig. 3.14 X-ray diffractograms recorded from the surface of specimens plasma-nitrided at different process conditions

Figure 3.13 shows the X-ray diffractograms recorded from the surface of 4330V specimens gas-nitrided/nitro-carburized under various conditions. For the specimens nitrided using the gas mixture of ammonia and air are oxidized as

Table 3.7 Comparison of some of the results obtained for plasma-nitrided, gas-nitrided, and nitro-carburized 4330V alloy specimens

Property	Plasma-nitriding	Gas-nitriding	Nitro-carburizing
White-layer thickness	Low (2–10 µm)	Intermediate (15–40 µm)	High (25–73 µm)
Penetration of white-layer	Low penetration <20 µm	Highest up to 244 µm	Intermediate up to 119 µm
Density of white-layer penetrated grain boundaries	Minimum	Maximum	Intermediate
Uniformity in white-layer	High	Less	Intermediate
Heat treatment cycle time (for a given case depth)	High (12–30 h)	Low (6 h)	Low (6 h)
Phases present	At low temperatures (<500 °C) and high r_N, ε-Fe$_3$N and γ'-Fe$_4$N are observed. However, at higher temperatures (>500 °C) and low r_N, γ'-Fe$_4$N is seen	Both ε-Fe$_3$N and γ'-Fe$_4$N are present, along with oxides (like, Fe$_2$O$_3$ and Fe$_3$O$_4$)	Only ε-Fe$_3$N phase is present since presence of carbon stabilizes the ε phase. Oxides are also evident
Temperature	Possible at temperatures as low as 460 °C	Temperatures needed are higher (>500 °C)	Temperatures needed are higher (>500 °C)
Case depth	Shorter case depth (100–150 µm) is achieved even with longer nitriding time (12–30 h)	Highest case depth (300 µm) is achieved just with nitriding time of 6 h	Case depth up to 200 µm can be achieved in 6 h
Hardness	Lower surface hardness (550–620 HV$_{0.1}$)	Higher surface hardness (690 HV$_{0.1}$)	Highest surface hardness due to presence of epsilon phase (750 HV$_{0.1}$)

evident from the presence of oxide peaks in the X-ray diffractogram. As mentioned above, the specimen (NC2) nitro-carburized at 540 °C shows the single phase compound-layer (Fig. 3.10) and X-ray diffractogram indicates the presence of ε-Fe_3N phase, along with traces of chromium nitrides (CrN) in the surface layer of the same specimen.

Referring to Fig. 3.14, it can be inferred that, at different nitriding temperatures, different phases are formed. From X-ray diffractogram it can be seen that at lower temperatures, viz., 460 °C near-surface region consists of ε-Fe_3N and γ'-Fe_4N phases, while at higher nitriding temperatures (>500 °C) near-surface region consists of γ'-Fe_4N phase (no indication for ε-Fe_3N phase). These results suggest that by selecting the appropriate nitriding temperature and gas ratio, the desired phases can be achieved in the white-layer. It is important to note that in Fig. 3.14, X-ray diffractogram recorded from the surface of specimen (Plasma 2) plasma-nitrided at 460 °C has a gas ratio of N_2:H_2::1:1, while for other two specimens (Plasma 1 and Plasma 3) the gas ratio is N_2:H_2::1:3. Therefore, the nitriding potential (r_N) for specimen Plasma 2 is higher than that of Plasma 1 and Plasma 3. This is also a potential reason for the formation of ε-Fe_3N phase near the surface region of Plasma 2. Referring to Lehrer's diagram [9, 10], this phenomenon which is dependent of nitriding temperature and potential can be explained. This diagram is useful for selecting the appropriate nitriding parameters (temperature and nitriding potential) as per the desired phases in the nitrided layer. Although this diagram is for pure Fe, which is in equilibrium with NH_3–H_2 gas mixture, it can be potentially used as a reference model for plasma-nitriding of Fe-based alloys.

3.4 Comparison of the Results

Table 3.7 summarizes all the results comparing all three treatment types (plasma-nitriding, gas-nitriding and nitro-carburizing), which will help in determining the most suitable process based on the property requirements.

3.5 Conclusions

- In case of plasma-nitrided 4330V alloy, different geometries of the specimen are subjected to different heating rates, and hence the kinetics of nitriding for the same component can differ at different locations. Different geometries can be subjected to the different current density, which causes nonhomogeneity in the temperature distribution and the nitriding case depth of the specimen.
- White-layer penetration and formation of iron-nitride network have been observed in the specimens irrespective of the type of processing technique, i.e., gas-/plasma-nitriding and nitro-carburizing. White-layer can penetrate deep in the diffusion zone even up to 244 µm and this penetration is the highest for

gas-nitrided specimen. However, this issue can be controlled by selecting the appropriate process parameters. Micrographs of annealed and gas-nitrided (at 550 °C using air:NH$_3$ gas ratio of 1:8) 4330V specimen have showed neither white-layer formation at the surface nor the nitride penetration inside the diffusion zone. However, the microhardness-depth profile confirmed the successful nitriding of the specimen.

- Since temperature is the dominant factor in any diffusional process, it controls the thickness of white-layer and total case depth. It has been observed that at higher temperatures, white-layer penetration and the density of iron-nitride network at corners are considerably higher compared to that of at lower temperatures. However, such behavior is not observed in the plasma-nitrided specimens.
- In case of plasma-nitriding, by appropriate selection of process parameters, a mono-phase compound-layer can be obtained. At higher nitriding temperatures, i.e., >500 °C, and lower N$_2$:H$_2$ gas ratio (1:3), the near-surface region consists of γ'-Fe$_4$N phase and no indication of ε-Fe$_3$N phase. Plasma-nitriding gave the lowest values of white-layer thickness compared to the gas-nitrided and nitro-carburized specimens.
- Highest values of surface hardness have observed for nitro-carburized specimen compared to the gas-/plasma-nitrided specimen.

Acknowledgment The author would like to thank Mr. Kunal Gokhale for his cooperation in preparing this chapter. The author would like to thank Bharat Forge Ltd., Pune and Bhat Metals Research Pvt. Ltd., Pune for plasma-nitriding experiments.

References

1. Pye D (2003) Practical nitriding and ferritic nitro carburizing. ASM International, Materials Park, Ohio, USA
2. Totten G (2007) Steel heat treatment handbook, 2nd edn. ASM International, Materials Park, Ohio, USA
3. Sun Y, Bell T (1997) Numerical model of plasma nitriding of low alloy steels. Mater Sci Eng 224A:33–47
4. Walkowicz J (2003) On the mechanisms of diode plasma nitriding in N$_2$-H$_2$ mixtures under DC- pulsed substrate biasing. Surf Coat Technol 174–175:1211–1219
5. Hosmani SS (2006) Nitriding of iron based alloys; role of excess nitrogen. PhD Thesis, Stuttgart University, Germany
6. Burakowski T (2004) Surface engineering of metals. In: Proceedings from 6th international tooling conference, vol 38, pp 413–520
7. Jeong B (2001) Effects of process parameters on layer formation behaviour of plasma nitrided steels. Surf Coat Technol 141:182–186
8. Czerwiec T, He H, Weber S, Dong C, Michel H (2006) On the occurrence of dual diffusion layers during plasma assisted nitriding of austenitic stainless steel. Surf Coat Technol 200:5289–5295
9. Mittemeijer EJ, Slycke JT (1996) Surf Eng 12:152
10. Mittemeijer EJ, Somers MAJ (1997) Surf Eng 13:483

Chapter 4
Recent Advances in Surface Alloying of Austenitic Stainless Steel by Plasma Nitriding

Abstract Surface alloying of materials by plasma nitriding using glow discharge plasma nitriding has become an important environmentally benign surface modification process to obtain improved hardness, wear resistance, and corrosion resistance. In this paper, surface alloying of austenitic stainless steels and Cr-coated austenitic stainless steel by plasma nitriding is discussed. Comparative studies on these materials are discussed in relation to kinetics, phase distribution, and nitriding mechanism. The chapter also deals with thermal stability, dimensional variation, wear resistance, and the mechanism of formation of Cr–N coatings.

Keywords Plasma nitriding · Austenitic stainless steel · Chrome-plated stainless steel · Diffusion · Microstructure · Abrasive wear

4.1 Introduction

Surface alloying of steels by conventional nitriding methods is an established industrial process and the resulting properties and microstructures are well documented [1–7]. The purpose of surface alloying with interstitial elements such as nitrogen, carbon, and boron is to increase wear resistance, fatigue resistance, and hardness without sacrificing the bulk properties [4–7]. Spies et al. [3] have reported an overview of the developments in the process control of gas nitriding and discuss the status of nitriding of nonferrous alloys.

In the past three decades, nitriding of steels using glow discharge or plasma or ion nitriding has become a powerful technique due to its advantages over other methods [8]. Plasma nitriding can be carried out at low substrate temperature in comparison with conventional methods because the chemical reactions that take place normally at very high temperatures can be achieved in a low temperature plasma state. Ion bombardment in combination with plasma-induced dissociation of the nitrogen–hydrogen gas mixture promotes chemisorption and surface layer

S. S. Hosmani et al., *An Introduction to Surface Alloying of Metals*,
SpringerBriefs in Manufacturing and Surface Engineering,
DOI: 10.1007/978-81-322-1889-0_4, © The Author(s) 2014

intermixing, resulting in diffusion coatings that can be formed at temperatures lower than in conventional gas nitriding. Since the process allows for lower workpiece temperature, distortion of the components can be minimized. The method has become an important environmentally benign surface structuring process as the technique does not require toxic cyanides, ammonia gas, etc.

Plasma nitriding has also been shown to accommodate other routes of surface modification, namely plasma carburizing, plasma boriding, plasma heating, etc. [9, 10]. Hossary et al. [11] carried out duplex treatment of AISI 304 austenitic stainless steel using radio frequency nitriding and DC magnetron sputtering of titanium and reported an increase in hardness by 1.4 times compared to nitrided ones. Schaaf et al. [12] reported laser nitriding by excimer laser in air and nitrogen gas at ambient pressure and found it to result in the formation of Fe–N austenite and ε-iron nitride in the laser nitrided surface. The method, in principle, can be applied to steels containing alloying elements in various proportions. In addition to steels such as nitralloy, Cr–Mo or Cr–Mo–V, ferritic, martensitic, and austenitic stainless steels, precipitation hardened stainless steels and hot work die steels, metals such as Nb, Ti, Zr, Si, and B could also be successfully plasma nitrided [13–18]. The growth behavior of plasma-nitrided workpieces made of type 316 with complex geometry was investigated by Alves et al. [19]. Significant differences in thickness and hardness of the resulting nitrided layers were observed as a function of the nitriding parameters. The distribution of charge density near the edges of those samples has been reported to affect the nitriding behavior.

In order to avoid loss of corrosion resistance of stainless steels due to precipitation of CrN in high temperature nitriding, low temperature plasma nitriding (<773 K) was carried out [20]. In the temperature range 523–773 K, nitrogen-enriched layers with significant increase in hardness were noticed due to the formation of an expanded austenite or "S phase" and the growth is comparable to plasma immersion ion implantation [21]. It was later suggested that the S-phase is thermodynamically a metastable phase and it would transform into stable phases of CrN when thermally annealed at certain temperatures for longer duration [22].

The time resolved in-situ X-ray diffraction (XRD), analysis of growing nitrided layers during plasma nitriding was conducted to obtain experimental data of growing compound layers for different plasma nitriding parameters. It was found that ε-phase occurred after nucleation of the γ'-Fe$_4$N phase [23]. In situ and real-time XRD analysis was also carried out using synchrotron radiation on type 304 stainless steel (SS) at 723 K [24]. Plasma nitriding of type 304 SS induced rapid development of the S-phase, which was metastable and was replaced by CrN and Fe$_4$N. The latter phases remained the dominant phases until the end of the long-term nitriding process.

In this work, plasma nitriding behavior of austenitic stainless steels (type 316 stainless steel and titanium-modified austenitic stainless steel also known as D-9 alloy) and chromium-plated type 316 stainless steel is investigated. The latter coating was developed for hard facing of grid plate sleeve components of fast reactors where resistance against galling of contacting surfaces, fretting, and corrosion by coolant fluid is of paramount importance. This study presents case depth, hardness, microstructure, and activation energy for the diffusion of nitrogen in nitrided austenitic stainless steels and chromium-coated type 316 stainless steel. The paper also highlights the thermal stability, dimensional changes, abrasive wear resistance, and nitriding mechanism of chromium-nitrided samples.

4.2 Issues in Nitriding of Austenitic Stainless Steels

Nitriding of austenitic stainless steel is drawing increasing attention due to its applications as construction materials in nuclear reactors, and in the chemical and food processing industries [25]. Although the corrosion resistance is excellent, their hardness and wear resistance is relatively low. As the presence of a tenacious chromium oxide layer on the surface of the stainless steel suppresses the nitriding efficiency of conventional methods [26], plasma nitriding is found to be an efficient method. Also, understanding the nitriding behavior of austenitic stainless steels is important because these steels differ from ferritic steels owing to variations in the concentration of alloying elements, solubility of nitrogen, and the crystal structure. While there are studies available on gas nitriding of austenitic stainless steels [1–6, 27, 28], the literature on plasma nitriding of austenitic stainless steels is limited [23, 24, 29–32].

It has been observed that conventional methods of surface alloying offer non-uniform coating on the surface partially because of the difficulty in removing the chromium oxide layer [27]. Additionally, formation of such layer cannot be fully eliminated during the gas nitriding process as the process is not carried out in vacuum. Although the hardness obtained in nitrided austenitic stainless steels is high (~ 900 VHN), there is also loss of corrosion resistance after surface alloying. Therefore, the corrosion properties of austenitic stainless steels have been improved by nitriding the samples at low substrate temperatures where there is no precipitation of CrN. Enhanced properties of both corrosion resistance and wear resistance were obtained by forming the S-phase formation (also known as γ_N in the literature) on the surface by reducing the treatment temperature [21–23, 33]. Besides these challenges, it is also important to retain the Cr content of the base alloy even after nitriding. However, formation of CrN or Cr_2N phases requires adequate supply of Cr to form nitrides. Formation of nitride phases at the surface of the components can lead to a loss of Cr on the surface and subsurface of the base alloy and results in the loss of mechanical properties and even transformation from γ to α phase locally [27]. Therefore, it is imperative to retain the Cr content of the base alloy while nitriding. There have been a few attempts to deposit Cr by electroplating followed by nitriding [13, 34].

Although electroplated chromium itself offers high hardness, corrosion resistance, and low coefficient of friction, there are limitations to the application of electroplated Cr coatings. The microcracks developed during electroplating reduce wear and corrosion resistance. Besides, the hardness decreases significantly with increasing temperature above 623 K [34]. In order to overcome these disadvantages, the surface of the electroplated chromium is modified by means of plasma nitriding. Chromium nitride coatings exhibit excellent thermal stability, wear, and corrosion resistance and have been found to be an alternative to titanium nitride coatings for tribological applications [35]. It has been reported that pure overlay coatings of CrN exhibit high hardness and poor toughness [36]. The toughness of these coatings can be enhanced by controlled addition of alloying elements such as Cu, Ni, etc. [37, 38]. However, formation of CrN in the matrix of austenitic stainless steel by surface alloying provides the required mechanical properties at the surface of the steel without any change in the bulk properties.

4.3 Process and Mechanisms

Gas nitriding became better understood after Lehrer proposed the temperature versus nitriding potential diagram [39]. The nitriding potential is given by the relationship between ammonia and hydrogen partial pressures. If a high nitriding potential is used during nitriding, many atoms of nitrogen will diffuse into the sample's surface to form compound layers of iron nitrides and a diffusion zone in the subsurface region in iron-based alloys. By contrast, complex reactions in the plasma and their unknown superposition determine the nitrogen activity of the plasma [40]. Direct observation of plasma reactions and correlation with materials properties is still difficult. However, once nitrogen is transferred from the plasma to the surface of the steel similar processes as in gas nitriding can be expected [23].

The application of a low pressure glow discharge nitriding was introduced by Berghause [41] and Bosse et al. [42] in 1932. In the plasma nitriding process, the material to be nitrided is used as the cathode in a glow discharge produced in a gas mixture that essentially contains nitrogen. The gas mixtures used are $N_2 + Ar$, $N_2 + H_2$, $NH_3 + Ar$. For rapid nitride growth, high power densities (0.1–10 mA/cm^2 and 0.5–1 keV) are desirable so that plasma nitriding is carried out in the abnormal glow discharge region of the I–V characteristic curve [43]. In spite of minor modifications in the process, such as nitriding in air environment [44] and in crossed magnetic field [45], the basic process remains the same. However, the process controls, arc suppression facility using pulsed power supply, and chamber design have been improved over time to nitride the engineering components of a variety of shapes of larger dimensions.

Emission spectroscopy of glow discharge in nitrogen shows the presence of N^+ and N_2^+ ions so that the cathode undergoes energetic ion bombardment besides absorbing neutral gas species such as atomic nitrogen. The identification of such ionic bombardment as the plasma nitriding mechanism is a major distinction which separates gas nitriding from plasma nitriding. Edhenhofer [46] and Lakhtin et al. [47] proposed a sputtering mechanism for iron (Fig. 4.1b) in which sputtered iron atoms react with nitrogen to form unstable FeN, which is backscattered onto the cathode and decomposes to lower nitrides such as Fe_2N, Fe_3N, and Fe_4N. A fraction of nitrogen atoms released due to such decomposition is believed to diffuse into steel. Hudis [48] studied the plasma nitriding behavior in a nitrogen–hydrogen–argon gas

Fig. 4.1 Schematic of the plasma nitriding mechanisms **a** nitrogen–hydrogen molecular ion mechanism, *I* N + H → NH⁺ (Combination), *II* NH⁺ → N + H + (Dissociation), *III* N Diffusion and **b** sputtering mechanism, *i* Sputtering of Fe, *ii* Formation of complex FeN, *iii* Redeposition, *iv* 4FeN → Fe₄N + 3N, *v* N Diffusion

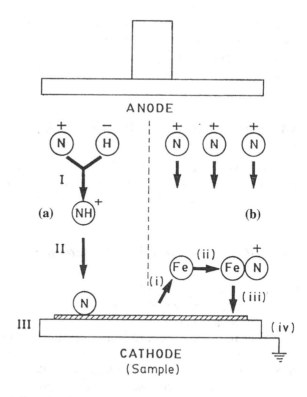

mixture with emphasis on the mechanism and the active plasma ingredients that cause nitriding. The cathode current mass spectrum as a function of gas composition indicates that in a nitrogen–hydrogen plasma, nitrogen ions (N^+, N_2^+, etc.) comprise less than 0.1 % of the cathode current while nitrogen–hydrogen molecular ions (NH_2^+, NH_3^+, etc.) comprise 10–20 %. These nitrogen–hydrogen molecular ions are capable of providing the required nitrogen upon hitting the cathode and offer higher hardness and case depth.

Brokman and Tuler [45] proposed a model for plasma nitriding based on the formation and diffusion of vacancy-nitrogen ion pairs for explaining accelerated case hardening associated with conventional nitriding. They performed nitriding in a crossed magnetic field to enhance the ion current density and found that diffusion coefficient was proportional to the current density. Subsequently, a collision-dissociation model [49] was proposed to account for possible sources of active nitrogen atoms in the N_2–H_2 gas mixture (Fig. 4.1a). The model supported the views of Hudis [48] and suggested that NH_j^+ ions and NH_j neutral atoms are produced as intermediate products and the total number of NH_j^+ ions were greater by a factor of 10 than the sum of N^+ and N_2^+ ions. This was attributed to the lower energy (<4.8 eV) required to produce dissociation of NH_j^+ ions into active nitrogen and hydrogen atoms.

The role of hydrogen as a constituent of the process gas mixture has also been identified [49]. The strong deoxidation of hydrogen atoms, a certain sputtering capability that helps to eliminate the passive film (e.g.) chromium oxide on stainless steel and activate the metal surface, and more importantly, the promotion of nitrogen–hydrogen molecular ions are some of the beneficial effects obtained in using hydrogen gas in the plasma nitriding process.

4.4 Surface Alloying by Plasma Nitriding

In the following sections, the experimental conditions used for plasma nitriding of austenitic stainless steels by plasma nitriding and chromium-plated austenitic stainless steel by pulsed plasma nitriding are briefly outlined.

4.4.1 DC Plasma Nitriding of Austenitic Stainless Steel

Solution-annealed type 316 stainless steel (17.0 wt%Cr, 11.9 wt%Ni, 2.25 wt%Mo, and 0.079 wt%C) samples of dimension $15 \times 10 \times 0.5$ mm^3, and titanium-modified stainless steel (15.09 wt%Cr, 15.037 wt%Ni, 0.21 wt%Ti, and 0.05 wt%C) samples of 10 mm diameter and 3 mm thickness were made and the surfaces to be nitrided were metallographically prepared. Prior to nitriding, the samples were ultrasonically cleaned in acetone and sputter cleaned in hydrogen atmosphere. The nitriding experiments were carried out in an apparatus shown in Fig. 4.2 using a direct current (DC) voltage of about 600 V; the details are discussed in [30, 31]. A low power radio frequency (r.f.) power supply (6 MHz) was also superposed to sustain the plasma. The stainless steel samples were nitrided in the temperature range 673–973 K for different amounts of time.

Recognizing the fact that direct plasma nitriding of type 316 stainless steel invariably results in nonuniform distribution of nitrides, a cyclic plasma nitriding process was evolved [31]. The process involves alternate plasma nitriding and diffusion annealing treatment. The process used low power of a few watts of r.f. discharge to produce nitrogen plasma and a DC power source to perform the nitriding process. During nitriding both the DC voltage and r.f. discharge were on, and during the diffusion part of the cycle, the DC power source was switched off. However, nitrogen plasma generated by the r.f. source was maintained during the diffusion cycle to prevent any nitrogen loss that may occur from the surface.

4.4.2 Pulsed Plasma Nitriding of Chromium-Plated Austenitic Stainless Steel

Conventional plasma nitriding that uses DC power supply encounters a limitation, such as the formation of undesirable cathodic arc, when intricate geometries or

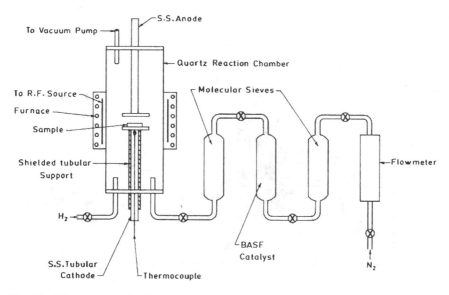

Fig. 4.2 A laboratory scale schematic diagram of plasma nitriding apparatus

surfaces with protrusions or oxide phases are treated. This is largely overcome with the pulsed DC power supply that can operate in the frequency range of 5–20 kHz. The use of pulsed sources enables the temperature of the specimen to be controlled by varying the width of the pulse without changing the bias voltage significantly. It has also been reported that the pulse width affects the microstructure and mechanical properties of the nitrided layers [50, 51]. The application of a pulsed DC power supply also has added advantages: (i) sputtering the substrate by positive ions is confined to the "ON" time of the duty cycle of the pulse, (ii) the active species responsible for nitriding remain active even during the "OFF" time of the pulse, (iii) rapid arcing resulting in the disruption of the nitriding process can be minimized, and (iv) it is possible to nitride narrow apertures by controlling the hollow cathode discharge [52–54], a phenomenon that gives rise to a special glow inside narrow apertures in the substrate. SS 316LN plate-shaped coupons (100 × 60 × 6 mm) were Cr plated up to 90 μm by standard electroplating process for the purpose of plasma nitriding [34].

The Cr platings in the as-deposited condition at room temperature had hardness of about 850–900 HV at a load of 100 g. The chromium-plated samples were ultrasonically cleaned in acetone and then loaded onto the cathode plate of the nitriding chamber. Pulsed plasma nitriding was carried out in an indigenously developed pulsed plasma nitriding facility. A schematic of a plasma nitriding system is shown in Fig. 4.3. The system essentially consists of a vacuum chamber, pumping system, pulsed DC power supply (900 V, 20 A, 1–20 kHz with variable duty cycle), mass flow controllers, and temperature controller. The pressure control, voltage control, and temperature control were performed through a personal

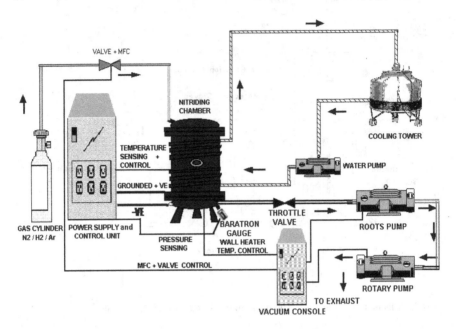

Fig. 4.3 A schematic of a pulsed plasma nitriding system

computer interfaced with the nitriding facility. Prior to plasma nitriding, sputter cleaning was conducted in an argon/hydrogen atmosphere at a pressure of about 2 mbar by biasing the samples negatively (500 V pulsed d.c.) at 523 K for 2 h. Subsequently, plasma nitriding was performed at 2–5 mbar of nitrogen and hydrogen gas mixture in the ratio 4:1 or 3:1 at a desired temperature for various durations of time [34]. The flow rates of nitrogen and hydrogen were monitored using mass flow controllers and the temperature was controlled by varying the duty cycle and hence the ion current. An auxiliary heater was used to nitride the sample at temperatures above 773 K.

4.4.3 Characterization Techniques

The phases present in the nitrided layer were analyzed by X-ray diffraction using a Philips PW 1730 X-ray diffractometer fitted with a graphite crystal monochromator. CuK$_\alpha$ radiation was used for these studies. Microhardness measurements were done using a Leitz microhardness tester at a load of 100 g. X-ray line profile analysis for the various elements in the nitrided zone was carried out on the edge preserved samples using a Philips PSEM-501 scanning electron microscope (SEM) in combination with EDAX 711B analyzer. Samples for transmission electron microscopy (TEM) were prepared from the nitrided foils by electrolytically polishing with 10 % perchloric acid in 90 % acetic acid and subsequently further thinned in a solution containing CrO_3 in 100 ml of orthophosphoric acid. The samples were examined in a Philips EM 400 HMG transmission electron microscope. Concentration profiles of nitrogen, iron, chromium, nickel, molybdenum, manganese, and titanium were determined using the Cameca SX-50 electron probe microanalyzer (EPMA). Abrasive wear tests of the coatings were carried out in air and at room temperature using a Calowear instrument (M/s CSEM Instruments SA, Switzerland).

4.5 Nitriding Behavior of Austenitic Stainless Steels

4.5.1 Structure, Microstructure, and Microhardness

Diffractometer traces obtained from the surface of the samples in the as-nitrided conditions are shown in Fig. 4.4. It is seen that samples nitrided at 723 or 848 K contain Fe_3N–Fe_2N and Fe_4N in large quantities, while precipitation of CrN is noticed at 973 K. The depth profiling of the nitrided layer obtained at 848 K, 25 h by XRD indicates that the surface layer consists of iron rich nitrides and the subsurface is constituted of CrN [31]. The thermal stability of these nitrides was investigated by aging the sample nitrided at 723 K, 26 h at 923 K various durations. It has been observed that at a temperature of 973 K, Fe_4N tends to dissolve in the austenite matrix [31].

Figure 4.5 shows a typical microstructure of the nitrided surface of type 316 stainless steel obtained on cyclic plasma nitriding at 848 K. The morphology of the surface indicates the uniform distribution of nitride particles of sizes in the range of 0.1–0.5 µm. The cyclic plasma nitriding when applied to type 316 stainless steel helps to dissociate the coarse nitride complexes (1–3 µm) on the surface [30] into finer ones and makes them diffuse further into the matrix during the diffusion annealing treatment. In addition to the other advantages, the cyclic plasma nitriding process leads to a uniform nitrided zone in type 316 stainless steel at temperatures less than 848 K where such uniformity is otherwise not achieved [31].

Transmission electron microscopy accompanied by microdiffraction technique was used to identify the microstructure of the nitrided layer and the region close to the interface between the matrix and the nitrided layer obtained in type 316 stainless steel. The globular or irregularly shaped particles were found to be γ'-Fe_4N phase (Fig. 4.6a). The lamellar aggregates were found to be CrN phase (Fig. 4.6b). The respective microdiffraction patterns are shown in the inset of these figures. The appearance of the lamellar colonies with respect to their spacing and sizes closely resembles discontinuous reactions. The matrix adjacent to the interface exhibits a mottled contrast [55]. One can positively infer the existence of a pre-precipitation region in which substitutional–interstitial solute

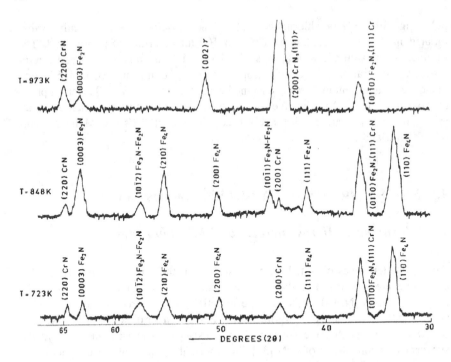

Fig. 4.4 Diffractometer traces of the cyclic plasma nitride samples at 723, 848 and 973 K for 25 h

Fig. 4.5 A typical microstructure of a uniformly nitride surface of type 316 stainless steel with cyclic plasma nitriding treatment at 848 K

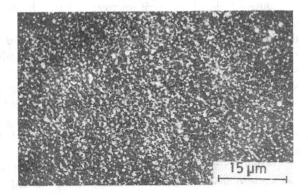

15 μm

atom clusters similar to Guiner-Preston (GP) zones in the matrix adjacent to the interface.

Figure 4.7a shows the microhardness-depth profile measurements across the nitrided zones in D-9 alloy nitrided in the temperature range 773–923 K for various durations at 848 K, 25 h. For a treatment temperature of 848 K, the hardness in the nitrided region is increased from a bulk value of 180 to 1,140 HV. The hardness remains constant throughout the layer and abruptly falls to a

Fig. 4.6 a TEM micrograph of a typical nitrided region (B.F. image, 823 K, 12 h). *Inset* shows the microdiffraction pattern of γ'-Fe4 N and **b** Lamellar colonies of γ and CrN (B.F. image, 823 K, 12 h). *Inset* shows the microdiffraction pattern of CrN

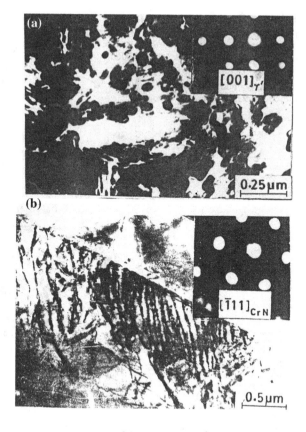

hardness value of 300 HV around 45 um from the surface. This indicates a sharp boundary between the nitrided layer and the bulk. The maximum hardness and the shape of the hardness-depth profiles obtained in nitrided D-9 alloy are similar to those obtained in plasma-nitrided type 316 stainless steel [31]. A typical microstructure of a specimen of D-9 alloy plasma nitrided at 923 K for 24 h is shown in Fig. 4.7b. It is observed that the nitrided layer is uniform and the layer boundary is parallel to the specimen surface, clearly indicating a planar growth front in accordance with hardness-depth profile measurements.

Strikingly, the D-9 steel that constitutes 15 wt% of Cr also shows a behavior similar to type 316 stainless steel [32]. The austenitic stainless steels in the present investigation contain Cr $\geq$ 15 wt% and these may be considered as a system in which a strong nitrogen–solute interaction prevails. Such a strong interaction results in the formation of a uniformly hard subsurface of CrN which advances progressively into the core causing a sharp boundary between the nitrided layer and the bulk. The effect of such a strong nitrogen–chromium interaction is seen in the hardness-depth profiles (Fig. 4.7a), where the hardness abruptly falls from a region of constant hardness to a region of bulk hardness.

Fig. 4.7 a Hardness-depth profiles of D-9 alloy plasma nitrided at 773, 848, and 923 K. and **b** SEM cross-sectional image of nitrided layer on D-9 alloy plasma nitrided at 923 K

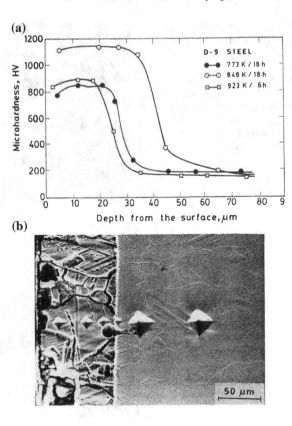

4.5.2 Internal Nitriding Model

It has been shown in this work that there are two types of nitrided layers while nitriding austenitic stainless steels: a surface layer of iron nitrides and an inner layer of chromium nitrides. Similar nitrided layers have been observed in some iron alloys [56]. In the case of nitrided D-9 alloy, the surface layer containing iron nitride is too thin (a few microns) and therefore, the kinetics of growth of the subsurface layer containing CrN is discussed according to the rate equation of internal nitriding model proposed by Jack and others [56, 57]. The validity and assumptions made while applying the model to D-9 steel is discussed in [32]. It is shown that the subsurface of the nitrided layer is controlled by the diffusional process of nitrogen in D-9 steel obeying a parabolic rate law, $X^2 = K_p t$, where X is the thickness of the subsurface layer containing CrN, t is the time of nitriding, and K_p is a rate constant. Figure 4.8a shows the square of the thickness, X^2 of the nitrided layer as a function of the nitriding time for the D-9 alloy nitrided at 773, 848, and 923 K, respectively. The square of the thickness of the nitrided layer is seen to increase linearly with time at each nitriding temperature. In addition, the

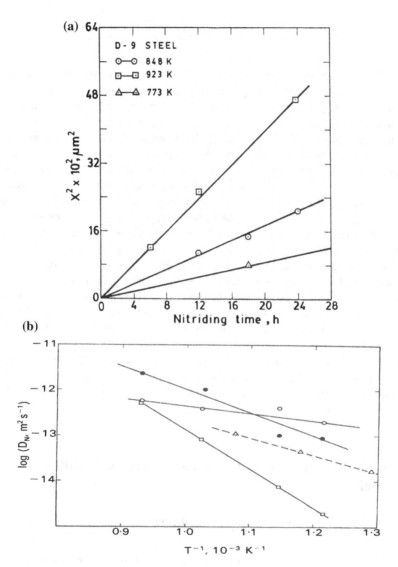

Fig. 4.8 a Square of thickness of nitrided layer (X^2) as a function of nitriding time for D-9 alloy nitrided at 773, 848, and 923 K, and **b** Plot of logDN versus reciprocal temperature for plasma nitrided D-9 alloy compared with corresponding plots for similar material from earlier work by Billon and Henry [27]

thickness of the nitrided layer increases with increase in temperature for a fixed treatment time because of the increased rate of nitrogen diffusion.

The internal nitriding model [56, 57] predicts that the subsurface-nitrided layer thickness (X) varies with time (t) according to

$$X^2 = \frac{2C_N D_N t}{C_{cr}\gamma} \qquad (4.1)$$

where C_N is the concentration of nitrogen at the surface of the internal nitrided layer (CrN), D_N is the diffusion coefficient of nitrogen in the nitrided layer, C_{Cr} is the concentration of the substitutional solute (Cr) in the nitrided layer, and γ is the atomic ratio of nitrogen to the solute element (Cr) in the nitrided phase.

Since the calculation of D_N involves the concentration of nitrogen and chromium in the internal nitriding layer, the quantitative analysis of the elements present in the nitrided layer for the temperature range 773–923 K for D-9 steel was carried out by EPMA. The concentration of nitrogen was found to increase sharply at the boundary between the nitrided layer and the bulk and thereafter remained more or less constant throughout the nitrided layer. There was more or less an abrupt linear fall of nitrogen concentration from the plateau region of the nitrided layer to the bulk region of zero concentration within a distance of 5 μm. The effect of such a strong nitrogen–chromium interaction was also seen in the shape of hardness profiles of austenitic stainless steels. The concentration profiles of nitrogen of this kind are essential in applying the internal nitriding approximation.

Equation (4.1) based on internal nitriding model was used to calculate the diffusion coefficients of nitrogen. The calculated value of D_N for D-9 steel at 773 K is 1.6×10^{-14} m^2/s, at 848 K it is 4.4×10^{-14} m^2/s, and at 923 K it is 1.04×10^{-13} m^2/s. A comparison of our data with those available in the literature [27, 58] is shown in Fig. 4.8b. The calculated activation energy for diffusion of nitrogen and the frequency factor are found to be 69.4 kJ/mol and 1.18×10^{-12} m^2/s, respectively.

4.6 Nitriding Behavior of Cr-Plated Austenitic Stainless Steel

4.6.1 Microstructure and Microhardness

Plasma nitriding of Cr-plated samples was carried out for various durations in the temperature range 833–1273 K. XRD analysis of the samples indicated the presence of hexagonal Cr$_2$N phase in the Cr matrix (Fig. 4.9). The intensity of the Cr$_2$N diffraction lines was found to increase with increasing nitriding temperature owing to increase in thickness of Cr$_2$N layer with temperature. The presence of untreated chromium was also observed to be below the surface layer of nitrides. Moreover, no trace of carbides and oxides was observed within the detectable limits of XRD. It is interesting to note that the nitrides were Cr$_2$N type and not CrN, which were reported in the plasma-nitrided-type 316 stainless steel [30, 31]. The formation of Cr$_2$N was found to be more dominant in preference to CrN at

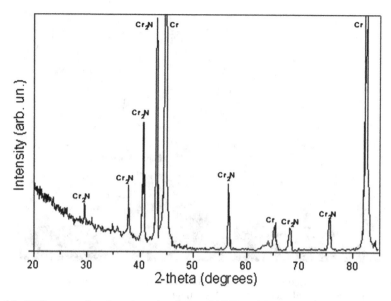

Fig. 4.9 XRD trace of a chromium-plated type 316LN stainless steel sample-pulsed plasma nitrided at 913 K, 45 h and gas flow ratio of N_2: H_2: 4:1

intermediate temperature ($\sim$ 900 K) and high temperature (873 K < T < 1273 K) nitriding of pure chromium.

Surface morphology of the nitrided samples showed a featureless structure with a random distribution of nitride particles [34]. Also, the surface was found to be etched due to ionic bombardment of nitrogen and hydrogen ions. A typical cross-sectional SEM microstructure of a sample of plasma nitrided chromium-plated stainless steel at 913 K, 45 h is shown in Fig. 4.10a. It is seen that the nitrided layer is uniform and the layer boundary is parallel to the sample surface, clearly indicating a planar growth front. Figure 4.10 also shows the indentation marks due to hardness measurements by a Vickers diamond pyramid indenter. The sizes of the indentation marks show a gradual decrease in size as a function of distance from the matrix to the nitrided layer. The hardness-depth profile (shown in the inset of Fig. 4.10b) reveals a uniform decrease in hardness as a function of depth for the samples nitrided at intermediate temperature of 913 K. The sample nitrided at 913 K, 45 h has a peak hardness of about 900 HV for a depth of about 10 μm, hardness plateau of about 700 HV for a depth of 60 μm, and finally a smoothly decreasing hardness profile. When the sample was nitrided at 913 K, 30 h, both the hardness and case depth were found to decrease. In both the cases, the decrease in hardness across the chromium-stainless steel interface is not abrupt. TEM study of plasma nitriding of chromium coating carried out at 993 K for 20 h indicated chromium nitride and carbide layers with thickness of 6–7 μm [59]. Interestingly, it was observed that after plasma nitriding the microcracks formed during electroplating were closed due to redeposition of ion sputtered materials, especially, chromium nitrides.

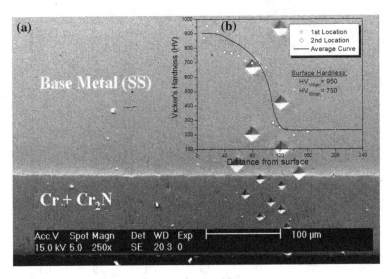

Fig. 4.10 **a** SEM cross-section of Cr–N coating on SS substrate showing microindentation marks and **b** hardness-depth profile across the Cr–N coating/SS

4.6.2 Kinetics of Nitriding

The observed case depths, hardness, and diffusion coefficient of nitrogen as a function of nitriding temperatures and time for plasma nitriding of chrome-plated stainless steel are shown in Table 4.1. The table shows that when the nitriding temperatures were less than 973 K, the case depths were small and the hardness obtained was less than 900 HV, while at temperatures more than 1073 K, there was considerable increase in the case depth and hardness exceeded 1,400 HV. Based on the time of nitriding (t) and case depth (X), the diffusion coefficients of nitrogen (D) were calculated from the relation, $X \approx 2\sqrt{(Dt)}$. These values are shown in Table 4.1. Figure 4.11 is lnD versus $1/T$ plot indicating that the nitrided layer growth is a diffusion-controlled process. The activation energy for nitrogen diffusion in chromium determined from the plot was found to be 131.4 kJ/mol [34]. This value is higher than the activation energy for diffusion of nitrogen in the nitrided layer of type 316 austenitic stainless steel (69.4 kJ/mol). The higher activation energy causes a significant reduction in the nitrided layer thickness in the Cr-plated stainless steel compared to austenitic stainless steels. It is suggested that the rate of diffusion of nitrogen is further restricted due to formation of a hard impervious Cr_2N phase at the surface of the Cr plating.

Table 4.1 Experimental conditions, case depths, hardness, and diffusion coefficients obtained in plasma nitriding of Cr-plated type 316 stainless steel

Temperature (K)	Time (h)	Case depth (μm)	Vickers hardness (HV)	Diffusion co-efficient (m²/s)
833	20	5	550	1.25×10^{-16}
973	3	6	850	3.33×10^{-15}
1073	3	15	1400	20.83×10^{-15}
1173	3	25	1650	57.87×10^{-15}
1273	3	40	1850	148.15×10^{-15}

Fig. 4.11 lnD versus 1/T plot for the plasma nitrided samples

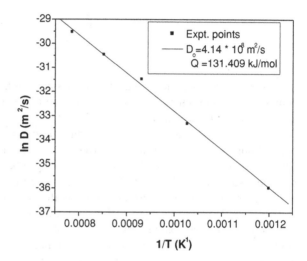

4.6.3 Thermal Stability, Distortion, and Wear Resistance

During electroplating of chromium, hydrogen that evolves from the solution gets penetrated into the sample and gets occluded in forming the deposit leading to high hardness of the coating. Hydrogen absorbed within Cr coating may also desorb from it, especially at higher service temperature. Therefore, the stability of the Cr coating as well as the stainless steel substrate was monitored by measuring the surface hardness as a function of nitriding temperature. For example, it was observed that at 823 K, the hardness decreased very slightly for stainless steel, while it decreased by three times for chromium coating [34]. However, the important advantage of plasma nitriding even at this temperature was that the surface hardness remained at about 850 HV due to the formation of chromium nitride phase. Studies on thermal stability of CrN and Cr_2N indicate that there was a decrease in hardness for CrN phase due to decomposition at temperature above 1,100 K [60]. On the other hand, we have observed Cr_2N even at temperature 1,273 K indicating the fact that Cr_2N is more stable than CrN and formation of such a phase is beneficial to the life of the component. While upgrading a

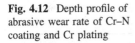

Fig. 4.12 Depth profile of abrasive wear rate of Cr–N coating and Cr plating

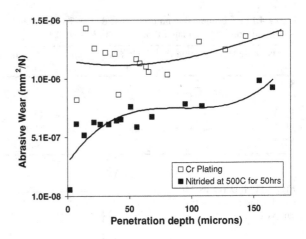

laboratory scale process to an industrial component, dimensional tolerance of large components assumes utmost importance. The percentage changes in dimensions of the tubular workpiece nitrided at 773 K, 50 h along the internal diameter (ID) which was chrome nitrided, and also along the outer diameter (OD) which was masked during the nitriding process. The distance of this distortion measurement was calculated from one end of a 970 mm long tube. It was observed that the average distortion was less than ± 0.15 % proving the fact that the pulsed plasma nitriding process does not introduce any significant distortion in the workpieces.

Figure 4.12 shows the depth profiles of the abrasive wear rates (in mm^2/N) for the chromium nitride coating processed by pulsed plasma nitriding at 773 K, 50 h compared to that of Cr coating. It is observed that the abrasive wear rate for Cr_2N coating is at least one order of magnitude lower than that for Cr coating. The abrasive wear rate is lower near the surface of the Cr_2N coating, and it gradually increases with depth of the coating. Beyond the coating/SS interface (i.e., $\sim$90 μm, the thickness of the coating), the wear rate increases sharply indicating the usefulness of coating on the SS substrate. On the other hand, the wear rate is nearly constant for Cr coating near the coating/SS interface, which increases in the SS substrate. Though the surface hardness of the Cr and Cr_2N coatings are nearly the same ($\sim$900 HV), Cr_2N is found to offer a superior abrasive wear property.

4.6.4 Nitriding Mechanism

Several theoretical models have been proposed to explain the kinetics of the process. Plasma species such as ions and energetic neutrals have a great effect on the kinetics of the nitriding process, since plasma nitriding times are generally shorter than gas nitriding. However, controversy still remains as to which plasma species are responsible for the nitriding process. For instance, Tibbetts [61] investigated the nitriding behavior of samples biased to repel positive ions and obtained rates similar to those of unbiased samples, leading to the conclusion that only energetic and neutral species are responsible for formation of a nitrided layer. Hudis et al. [48] suggested that the mass transfer from N_2^+ and NH_2^+ species is the predominant but not unique mechanism in ion nitriding. It was further demonstrated by Gantois et al. [62] and Alves et al. [51] that the post discharge region of the pulsed plasma nitriding contains only the excited neutral species and atomic nitrogen.

A nitriding mechanism for formation of Cr_2N by plasma nitriding was proposed in accordance with the classical works on plasma nitriding of steels [30, 31, 63] and based on the formation of Cr_2N. Transfer of nitrogen from the plasma into the surface of Cr layer could occur by direct nitrogen ion occlusion by chemisorption. In addition, the atomic nitrogen (neutrals) in the plasma may also react directly with sputtered Cr forming CrN and Cr_2N. Since CrN has not been observed in the XRD results shown in Fig. 4.4, it is believed that they react in such a way as to produce Cr_2N and N. For the sake of simplicity, we assume here that N neutrals are produced by cracking of N_2 gas. The atomic nitrogen thus released could diffuse into the Cr coating. Accordingly, the various steps leading to the diffusion of nitrogen into the bulk of Cr layer are given below:

$$Cr + N \rightarrow CrN \text{ (adsorbed); and } 2Cr + N \rightarrow Cr_2N \text{ (adsorbed)}$$
$$2\,CrN \rightarrow Cr_2N + N \text{ (diffusing species)}$$

Additionally, the stable Cr_2N layer that forms first on the surface of the coating possibly also retards the direct diffusion of nitrogen and hence slows down the diffusion kinetics resulting in long durations of nitriding. From this point of view, the process of indirect diffusion of nitrogen through reactions in metastable Cr–N and subsequent formation of Cr_2N appears to be the dominant mechanism for N transfer. The diffusion of this indirectly generated nitrogen is time–temperature-dependent. Since no studies are available for the formation of Cr_2N in plasma nitriding of austenitic stainless steel, the formation of such phase is well promoted while nitriding pure Cr or steels with high Cr content. Since a graded hardness profile has been observed (Fig. 4.10b), it is also plausible that the nitrogen concentration profile might exhibit a graded profile. This mechanism of indirect diffusion of nitrogen through reactions in metastable chromium nitrides by virtue of plasma nitriding offers a major advantage over other thermochemical processes such as gas nitriding.

4.7 Conclusions

The plasma nitriding behavior of austenitic stainless steels and chrome-plated austenitic stainless steel is investigated in this work. Unlike the conventional gas nitriding process, plasma used in the nitriding process enhances the nitriding behavior and modifies the nitriding mechanism significantly. The mechanism of nitriding in these steels was discussed qualitatively to understand the microstructures obtained in these materials after surface alloying by nitriding. Some important conclusions are listed below:

(i) In type 316 stainless steel, an Fe-rich surface zone consisting of mostly iron nitrides followed by a subsurface zone of CrN precipitation was obtained. D-9 steel had a thin surface layer of γ'-Fe$_4$N followed by subsurface layer of CrN. Both the materials indicated similar nitriding behavior after surface alloying by direct current plasma nitriding.

(ii) Both type 316 stainless steel and D-9 alloy showed a maximum surface hardness of about 1,100 HV and the hardness-depth profiles showed a sharp interface between the nitride case and the matrix. Since these steels contain Cr $\geq$ 15 wt% they may be considered as a system in which a strong nitrogen–solute interaction prevails. Such a strong interaction results in the formation of a uniformly hard subsurface of CrN which advances progressively into the core causing a sharp boundary between the nitrided layer and the bulk.

(iii) From the concentration-depth profiles of nitrogen in D-9 steel, the diffusion coefficients of nitrogen in the nitrided layer in the temperature range 773–923 K and the activation energy for the diffusion of nitrogen were calculated using the rate equation of internal nitriding model.

(iv) In contrast to the nitriding behavior of austenitic stainless steel, plasma-nitrided chrome-plated austenitic stainless steel showed that the coating is a mixture of polycrystalline Cr$_2$N and Cr after surface alloying.

(v) Surface hardness of the coating is about 950 VHN. Hardness was found to decrease gradually with depth. The abrasive wear rate of the coating was nearly an order superior to that for Cr coating and the dimensional changes in the chrome-nitrided component were less than 0.01 %.

(vi) The conversion of Cr into CrN occurs by nitrogen diffusion, the slow kinetics of which necessitates a long process duration. A mechanism for nitrogen diffusion in Cr-coating is proposed on the basis of direct and indirect transfer of nitrogen into Cr coating. Indirect transfer of nitrogen could possibly occur due to electron–gas interactions through the decomposition metastable CrN to form Cr$_2$N.

Acknowledgments The author (P. Kuppusami) gratefully acknowledges the support and encouragement given by Dr. V.S. Raghunathan, former Associate Director, Mr. A. L.E. Terrance and Dr. Arup Dasgupta, IGCAR, Kalpakkam for the contribution and support.

References

1. Mittemeijer EJ, Vogels ABP, Va Der Schaaf PJ (1982) J Mater Sci 15:3129–3140
2. Mirdha S, Jack DJ (1982) Metal Sci 16:398–404
3. Spies HJ, Reinhold B, Wilsdorf K (2001) Surf Eng 17:41–47

4. Karamis MB (1991) Thin Solid Films 203:49
5. Li CX, Sun Y, Bell T (2000) J Mater Sci Lett 19:1793
6. Karaoglu S (2003) Mater Charact 49:1793
7. Staines T (1990) Heat Treat Met 4:85
8. Jones CK, Martin S, Sturges DJ, Hudis M (1973) Heat treatment, vol 73. Metals Society, London, p 71
9. Jewbury P (1986) Mater Forum 9(3):179–181
10. Rie KT, Menthe E, Mathews A, Legg K, Chin J (1996) MRS Bull 21:46–51
11. El-Hossary M, Negam NZ, El-Rahman AM, Hammad M, Templier C (2008) Surf Coat Technol 202:1392–1400
12. Schaaf P, Emmel A, Illgner C, Lieb KP, Schubert E, Bergmann HW (1995) Mater Sci Eng A 197:L1–L4
13. Menthe E, Rie KT (1999) Surf Coat Technol 112:217
14. Lunarska E, Nikiforow K, Wierzchon T, Pokorska U (2001) Surf Coat Technol 145:139
15. Rolinski E, Sharp G, Cowgill DF, Peterman DJ (1998) J Nucl Mater 252:200–208
16. Gredelj S, Gerson AR, Kumar S, McIntyre NS (2002) Appl Surf Sci 199:234–247
17. Visuttipitukul P, Aizawa T, Kuwahara H (2003) Mater Trans 44(7):1412–1418
18. Pinasco MR, Ienco MG, GurNone P, Bocchini GF (2000) J Mater Sci 35:4079–4086
19. Alves C Jr, da Silva EF, Martinelli AE (2001) Surf Coat Technol 139:1–5
20. Larisch B, Brusky U, Spies H (1999) Surf Coat Technol 116–119:205–211
21. Collins GA, Hutchings R, Tendys J (1993) Surf Coat Technol 59:267–273
22. Li XY (2001) Surf Eng 17(2):147–152
23. Hirsch T, Clarke TGR, da Silva Rocha A (2007) Surf Coat Technol 201:6380–6386
24. Feugeas B, Gomez AC (2002) Surf Coat Technol 157:167–175
25. Metals Handbook, 9th Edition (1981) Heat treating. American Society for Metals, Materials Park, Ohio, 4:211
26. Takada J, Ohizumi Y, Miyamura H, Kuwahara H, Kikuchi S, Tamura I (1986) J Mater Sci 21:2493–2496
27. Billon B, Hendry A (1985) Surf Eng 1(2):125–130
28. Chung MF, Yap AK, Lim YK (1985) Scr Metall 19:415–419
29. Chung MF, Lim YK (1986) Scr Metall 20:807
30. Sundararaman D, Kuppusami P, Raghunathan VS (1983) Surf Technol 18:341–347
31. Sundararaman D, Kuppusami P, Raghunathan VS (1987) Surf Coat Technol 30:343
32. Kuppusami P, Terrance ALE, Sundararaman D, Raghunathan VS (1993) Surf Eng 9:142
33. Zhang ZL, Bell T (1985) Surf Eng 1:131
34. Kuppusami P, Dasgupta A, Raghunathan VS (2002) J Iron Steel Inst Japan Int 42:1457–1460
35. Pierson H (1996) Handbook of refractory carbides and nitrides. Noyes Publications, New Jersey, p 18
36. Elangovan T, Kuppusami P, Thirumurugesan R, Ganesan V, Mohandas E, Mangalaraj D (2010) Mater Sci Eng, B 167:17–25
37. Kuppusami P, Elangovan T, Murugesan S, Thirumurugesan R, Khan R, Ramaseshan R, Divakar R, Mohandas E, Mangalaraj D (2012) Surf Eng 28(2):134–140
38. Musil J, Jirout M (2007) Surf Coat Technol 201:5148–5152
39. Lehrer B (1930) Z Elektrochem 6(6):383
40. Kooi BJ, Somers MA, Mittemeijer EJ (1996) Metall Mater Trans A Phys Metall. Mater Sci 27A:1063
41. Berghause B (1932) German Patent 669:639
42. Von Bosse J et al (1932) Swiss Patent 172:432
43. Venugopalan M, Asni R (1985) In: Klabunde KJ (ed) Thin films from free atoms and particles. Academic Press, Inc. Orlando, p 54
44. Brokman A (1980) J Vac Tech 17:657
45. Brokman A, Tyler FR (1981) J Appl Phys 52(1):468
46. Edenhofer B (1974) Heat Treat Met 1:23
47. Lakhtin YM, Kogan YD, Saposhnikov VN (1976) Metalloved Therm Obrab Met 6:2

48. Hudis M (1973) J Appl Phys 44(4):1489
49. Bingzhang X, Yingzhi Z (1987) Surf Eng 3(3):226
50. Rie KT (1989) In: Proceedings of international conference ion nitriding-carburising. Materials Park, ASM International, Cincinnati, OH, p 45
51. Alves C, Rodriques JA, Martinelli AE (1999) Surf Coat Technol 116–119:112
52. Kim YM, Han JG (2003) Surf Coat Technol 171:205
53. Gruen R, Guenther H (1991) Mater Sci Eng A 140 (1991)
54. Hugon R, Fabry N, Herrion GJ (1996) Phys D: Appl Phys 29:761
55. Kuppusami P, Sundararaman D, Raghunathan VS (1987) In: Proceedings of second International Congress on surface engineering, Stratford upon Avon, paper no. 31
56. Jack KH (1975) In: Proceedings of heat treatment. The Metals Society, London, Dec 1973, p 39
57. Lightfoot BJ, Jack DH (1973) Heat treatment, vol 73. Metals Society, London, pp 248–254
58. Grieveson P, Turkdogan ET (1964) Trans AIME 230:407
59. Wang L, Nam KS, Kwon SC (2003) Appl Surf Sci 207:372–377
60. Heau C, Fillit RY, Vaux F, Pascaretti F (1999) Surf Coat Technol 120–121:200
61. Tibbetts GG (1974) J Appl Phys 45:5072
62. Gantois M, Ablitzer D, Marchand JL, Michel H (1998) In: Proceedings of International Conference Heat Treatments of Materials, Materials Park. ASM International, Cincinnati, OH, p 55–66
63. Rizk AS, McCulloch DJ (1979) Surf Technol 9:303

Chapter 5
Chromizing, Carburizing, and Duplex Surface Treatment

Abstract The effect of different types of surface treatment on microstructure and hardness of mild steel was studied. The chosen surface treatments were carburizing, chromizing, and duplex surface treatment. Duplex surface treatment involved carburizing-then-chromizing and chromizing-then-carburizing. Surface-treated specimens were characterized using optical microscope, scanning electron microscope, X-ray diffraction, and microhardness measurements. The improvement in surface-hardness of the specimens was significant for chromized specimen. Duplex surface-treated specimens showed lower surface hardness than the chromized specimen, but it was more than the carburized specimen. Carbon-depleted zone was formed beneath the chromium-carbides layer for the chromized and carburized-then-chromized specimens which leads to undesirable hardness variation across the cross-section of the specimens, i.e., soft layer embedded between harder surface layer and unaffected core. Such hardness variation was not observed for the cross-section of chromized-then-carburized specimens and therefore, this sequence in the duplex treatment created a gradual transition in properties from surface to nontreated core of the cross-section.

Keywords Mild-steel · Chromium-carbide · Hardness

5.1 Introduction

Performance of materials in many applications can be enhanced by improving their surface properties. The inherent poor surface properties (mechanical, corrosion, wear properties) of some iron-based alloys have been a barrier to their wider applications in the automobile, food, chemical, petrochemical, paper, nuclear, and medical sectors. There are many methods for surface engineering of ferrous alloy components, such as surface alloying (using techniques like, pack, gaseous, plasma, ion beam, and salt-bath), surface hardening heat-treatment (using techniques like, induction, flame, laser/electron beam), and surface coatings. Using an

S. S. Hosmani et al., *An Introduction to Surface Alloying of Metals*,
SpringerBriefs in Manufacturing and Surface Engineering,
DOI: 10.1007/978-81-322-1889-0_5, © The Author(s) 2014

effective surface treatment, less expensive grades of steels (with lower alloy content, or with less expensive alloying elements substituting the more expensive ones) can possibly be used for comparable or even improved service life and performance.

In this paper, thermochemical surface treatments, like carburizing, chromizing, and duplex surface treatment (combination of carburizing and chromizing), of mild steel was carried out using pack-method. Carburizing involves the diffusion of carbon into the surface resulting in hard surface-layer [1, 2]. Apart from the hardness and wear resistance, chromizing enhances the corrosion and high temperature oxidation resistance of steel [3–7]. Diffusion of chromium is lower than the diffusion of carbon in steel at any given temperature of the solid phase due to the larger size of the chromium atom than carbon atom. This leads to the smaller case-depth for chromizing than carburizing at constant temperature and time. However, surface mechanical attrition treatment of the steel showed a much thicker chromium-diffusion layer than the coarse-grained steel after the same chromizing treatment, especially at low temperatures [4]. Combination of the properties which are unobtainable through any individual surface technology can be obtained by two (or more) established surface technologies, i.e., duplex surface engineering [8]. Duplex surface engineering, as the name implies, involves the sequential application of two (or more) established surface technologies to produce a surface composite with combined properties, which are unobtainable through any individual surface technology [8]. Such duplex treatments are promising in improving the surface properties of alloys, for example (i) steels can be duplex treated using carburizing and nitriding [9] or ceramic coating and nitriding [8] or nitrocarburizing and low temperature chromizing [10] or plasma nitriding and nickel/diamond coating [11] and (ii) titanium alloys can be duplex treated using diamond-like coating and oxidation [8, 12]. Along with wear property, antibacterial property of AISI 304 steel can be improved by duplex treatment of plasma alloying with copper on the plasma alloyed steel with nickel [13].

Against this background, this work is devoted to study the effect of the *sequence* of treatments in duplex surface treatment on the microstructure and properties, like hardness of steel. In this regard, steel specimens were carburized-then-chromized and chromized-then-carburized.

5.2 Experimental

Cylindrical-shaped specimens of mild steel were used for the surface treatments. The composition of mild-steel is as follows: C: 0.30 wt%, Mn: 0.55 wt%, Si: 0.90 wt%, S < 0.04 wt%, and P < 0.04 wt%. Grinding, polishing, and cleaning of the specimens were done before the treatment to remove rust and other contaminants. Carburizing was done at 1,000 °C for 5 h using the pack mixture of activated charcoal (85 wt%), $BaCO_3$ (10 wt%), $CaCO_3$ (2 wt%), and Na_2CO_3 (3 wt%). Chromizing was done at 1,000 °C for 5 h using the pack mixture of chromium powder (50 wt%), Al_2O_3 (47 wt%), and NH_4Cl (3 wt%). Following two types of duplex surface treatments were performed on mild steel: (ii) carburizing followed by chromizing and (ii) chromizing followed by carburizing. In the following sections, these duplex treatments are designated as "carburizing-then-chromizing" and "chromizing-then-carburizing," respectively. Carburizing and chromizing process-parameters used here are the same as mentioned above.

Cross-sections of the various surface-treated specimens were analyzed using scanning electron microscope (SEM), energy dispersive spectroscopy (EDS), and microhardness tester. The SEM micrographs and EDS measurements were taken with a JOEL-JSM-5900LV operating at 20 kV. The microhardness measurements were obtained by using a Vickers diamond pyramid indenter, applying a load of 300 g. X-ray diffractograms were recorded from the specimen surface using CuK_α radiation. The scanned diffraction angle (2θ) range was 30–100°. PCPDFWIN database was used to identify the phases present in the surface-treated steel specimens.

5.3 Results

Micrographs, obtained by SEM, of the cross-section of nontreated, carburized, and chromized mild steel specimens are shown in Fig. 5.1. All micrographs are at the same magnification. Surface of the cross-section is on the right side of each micrograph with increasing depth in the left direction. Colonies of pearlite, a lamellar structure, are present in all microstructures. Area-fraction of the pearlite in the micrograph of carburized specimen increases (Fig. 5.1b), which indicates the increase in carbon content due to the carburizing treatment. However, a small region near to the surface of the carburized specimen is free from pearlite colonies (Fig. 5.1b), which indicates that carbon was lost from the surface, i.e., decarburized layer forms. Decarburization is a frequently encountered problem during heat-treatment of steel components. This problem occurs due to the oxidizing atmosphere that reacts with carbon present in the steel surface region to produce CO or CO_2. The carburizing solid pack-mixture is not capable of generating enough carburizing potential surrounding the specimen below 800 °C [14]. Therefore, during cooling of the specimen after carburizing treatment the chemical potential of carbon in the surface region of specimen is more than that of in the surrounding atmosphere and this is the favorable condition for "decarburization." Micrographs of the chromized specimen (Figs. 5.1c and 5.2a) show the formation of chromized layer at the surface (which is further confirmed by the high hardness: see Fig. 5.6b), and a considerably thicker layer beneath it is a "carbon-depleted region" (with a thickness of about 50–80 μm), as indicated by the absence of pearlite colonies. Below this carbon-depleted region, the unaffected core is present which appears similar to Fig. 5.1a.

Microstructure of the cross-section of duplex surface-treated mild steel specimens is shown in Figs. 5.2b–c and 5.3. Micrograph shown in Figs. 5.2b and 5.3a

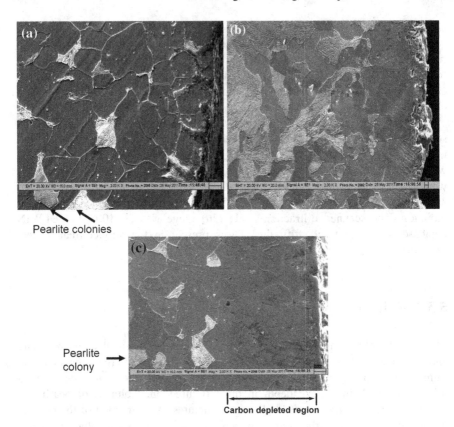

Fig. 5.1 Scanning electron micrographs of etched cross-section of **a** non-treated **b** carburized and **c** chromized mild steel specimens. The specimens were carburized and chromized at 1,000 °C for 5 h

corresponds to the carburized-then-chromized specimen while Figs. 5.2c and 5.3a correspond to the chromized-then-carburized specimen. Right portion of Figs. 5.2c and 5.3a has similar features to that of Figs. 5.1c and 5.2a, i.e., the presence of "carbon-depleted region" below the chromized layer. Here, the carbon-depleted region is embedded between chromized layer and the deeper region of already carburized layer. However, the pearlite colonies are observed below the chromized layer of the chromized-then-carburized specimen (Figs. 5.2c and 5.3b). In other words, the "carbon-depleted region" is *not* observed for the chromized-then-carburized specimen.

SEM-EDS results of the chromized-then-carburized specimen are summarized in Fig. 5.4. Energy dispersive spectroscopy spectrums were recorded at two locations, one within the chromized layer and other within the carburized layer. These locations are marked as location-1 and 2, respectively in Fig. 5.4a. The presence of Cr, Fe, and C at location-1 (Fig. 5.4b) and the presence of Fe and C at location-2 (Fig. 5.4c) confirm the surface layer is "chromized layer."

Fig. 5.2 Optical micrographs of etched cross-section of **a** chromized **b** carburized-then-chromized and **c** chromized-then-carburized mild steel specimens. The specimens were chromized and carburized at 1,000 °C for 5 h. (Optical microscope magnification: 500X)

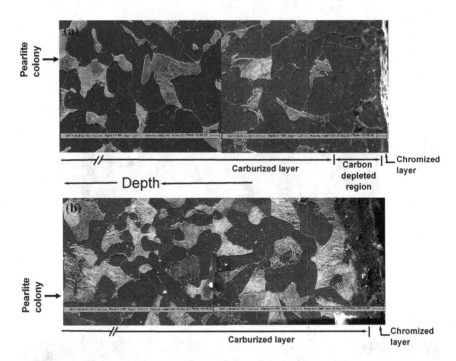

Fig. 5.3 Scanning electron micrographs of etched cross-section of **a** carburized-then-chromized and **b** chromized-then-carburized mild steel specimens. The specimens were carburized and chromized at 1,000 °C for 5 h

Concentration-depth profiles of Cr and Fe along the line-1, indicated in Fig. 5.4a, are shown qualitatively in Fig. 5.4c. It must be noted here that the concentration of the element, like carbon, measured by EDS (coupled with SEM) may not be considered reliable in practice and therefore, its concentration is not measured with depth. The concentration of Cr within the chromized layer decreases sharply with depth.

X-ray diffractograms, recorded from the surface of treated specimens, are shown in Fig. 5.5. X-ray diffractogram of nontreated specimen shows the presence of α-Fe. Due to the low carbon content in the nontreated specimen, cementite (Fe_3C) content is too low (Fig. 5.1a) to give the X-ray diffraction peaks of sufficiently high intensity. X-ray diffractogram of the carburized specimen reveal the presence of α-Fe and Fe_3C. X-ray diffractograms of chromized, carburized-then-chromized and chromized-then-carburized specimens show the presence of chromium-carbides, like $Cr_{23}C_6$, Cr_2C, and Cr_3C_2. Chromized and carburized-then-chromized specimens show also the small intensity X-ray diffraction peaks of Cr. Additional peaks of Fe_3C are present in the X-ray diffractogram of carburized-then-chromized specimen.

Microhardness-depth profiles of the cross-section of carburized, chromized, carburized-then-chromized and chromized-then-carburized specimens are shown in Fig. 5.6. Hardness of the carburized specimen decreases continuously with

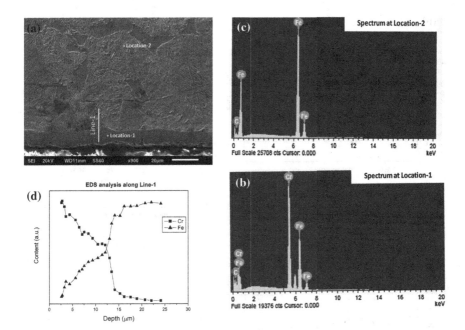

Fig. 5.4 a Scanning electron micrographs of etched cross-section of chromized-then-carburized mild steel specimens indicating the locations (location 1, 2 and line-1) of EDS measurements. Energy dispersive spectroscopy spectrum recorded from location-1 (**b**) and location-2 (**c**). **d** Concentration-depth profiles of Cr and Fe along the line-1

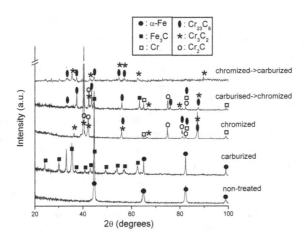

Fig. 5.5 X-ray diffractograms recorded from the surface of the mild steel specimens which are subjected to various surface treatments. The sequence of X-ray diffractograms from bottom to top is for the nontreated, carburized, chromized, carburized-then-chromized and chromized-then-carburized specimens. The specimens were carburized and/or chromized at 1,000 °C for 5 h

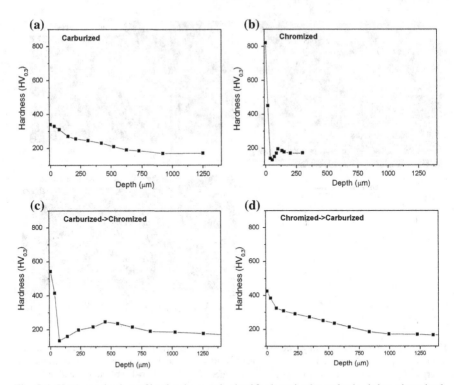

Fig. 5.6 Hardness-depth profiles for the **a** carburized **b** chromized **c** carburized-then-chromized and **d** chromized-then-carburized mild steel specimens. The specimens were carburized and chromized at 1,000 °C for 5 h

increase in the depth. Hardness of the chromized layer is the highest among the various treated specimens. Hardness decreases to the lowest value in the region beneath the chromized layer. Low hardness of this region between chromized-layer and nontreated core is due to the presence of "carbon-depleted region" (cf. Fig. 5.1c). Similar behavior is observed in the carburized-then-chromized specimen. The fluctuations of the hardness-depth profile are shown in Fig. 5.6c are associated with the chromized layer (high hardness region near to the surface), carbon-depleted region (low hardness region beneath the chromized layer), end-portion of the carburized layer (hardness gradient region from about 500 μm depth), and nontreated core. However, the chromized-then-carburized specimen shows the continuous decease in hardness with depth until the nontreated core is reached. Steeper hardness gradient near to the specimen surface becomes less steep after 70 μm depth. This is due to the presence of hard chromium-carbide layer near the surface and subsequent carburized layer. Width of the steeper hardness-gradient region in chromized-then-carburized specimen is bigger than the chromized specimen (70 vs. 40 μm).

Table 5.1 Surface hardness value of nontreated and various surface-treated mild steel specimens

	Non-treated	Carburized	Chromized	Carburized-then-chromized	Chromized-then-carburized
Surface hardness ($HV_{0.3}$)	170 ± 10	310 ± 20	800 ± 35	520 ± 20	410 ± 15

The specimens were carburized and/or chromized at 1,000 °C for 5 h

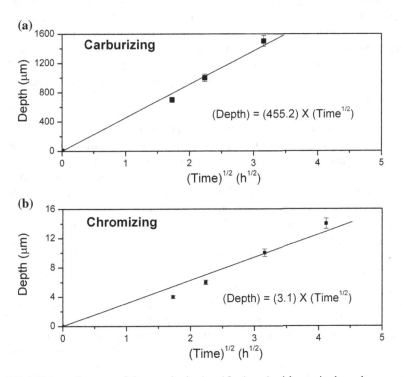

Fig. 5.7 Thickness increase of the **a** carburized and **b** chromized layers in dependence on the treatment time at 1,000 °C

Microhardness value at the *surface* of non-treated and various surface-treated specimens are shown in Table 5.1. Maximum surface hardness is observed for the chromized specimen. Duplex surface-treated specimens have lower surface hardness than the chromized specimen, but higher than the carburized specimen.

The determination of the thickness of the carburized and chromized layers was performed using optical micrographs of the specimen cross-sections. The thickness increase of the carburized and chromized layers in dependence on the treatment time at 1,000 °C is shown in Fig. 5.7. The layer thickness increases as a function of time for both treatments. However, the increase in thickness for carburizing is more pronounced than for the chromizing. This confirms the known fact of higher diffusivity of carbon than chromium at a constant temperature.

5.4 Discussion

The microscopical observations, supported by X-ray diffraction and microhardness measurements, can be interpreted as follows. During chromizing of mild steel at high temperature (1,000 °C), austenite (i.e., face-centered-cubic iron) phase, with dissolved carbon, is formed throughout the specimen. Chromium from the gas phase, formed in the chromizing pack-mixture, is deposited on the steel surface and starts diffusing into the austenite. Carbon has strong affinity toward chromium. Due to the smaller size of the carbon atoms, they can move relatively freely through the austenite toward the chromium coating and chromium-carbide formation occurs. Consumption of carbon by chromium coating, to form the hard carbides, leads to the formation of "carbon-depleted region." Similar reason can describe the formation of carbon-depleted region in the carburized-then-chromized specimen. However, in case of the chromized-then-carburized specimen, carbon-depleted region is filled with carbon during post-chromizing treatment, i.e., carburizing. Therefore, the pearlite is present just below the hard chromized layer. Hard chromium-carbide layer followed by decreasing carbon content with depth cause the smooth hardness-gradient up to the nontreated core of the chromized-then-carburized specimen. The mechanism of layer formation in carburized-then-chromized and chromized-then-carburized specimens is shown in the form of schematic diagram in Figs. 5.8 and 5.9, respectively, where numbers indicate the possible sequence of the events. The presence of chromium-carbides of various stoichiometry and the mechanism of layer development, as discussed above, suggest that chromium-rich carbide ($Cr_{23}C_6$) forms at the surface and carbon-rich carbide (Cr_3C_2) forms at the bottom of chromized-layer.

Lower surface-hardness of duplex surface-treated specimens than the chromized specimen can be interpreted as follows. During carburizing at 1,000 °C, carbon concentration depth-profile develops with surface carbon content of about 1.8 wt% [14, 15] and proeutectoid cementite forms along the pearlite grain-boundaries. At 1,000 °C, chromizing of such carburized specimen is unable to convert the carburized microstructure (containing ferrite plus cementite) near to the specimen surface into fully austenite phase, and therefore, cementite remains in the final microstructure after chromizing (confirmed the presence of cementite peaks in the X-ray diffractogram of carburized-then-chromized specimen: see Fig. 5.5). Cementite has lower hardness than the chromium-carbides [16] and, therefore, its presence in chromium-carbide layer causes decrease in the hardness. In case of chromized-then-carburized specimen, X-ray diffractogram is not identical to the X-ray diffractogram of chromized specimen, which indicates that the chromized layer becomes unstable and grows further during carburizing at 1,000 °C. The layer growth is confirmed by the larger width of the steeper hardness-gradient region in chromized-then-carburized specimen than in the chromized specimen (70 vs. 40 μm).

Tendency for the formation of "carbon-depleted region" in the chromized and carburized-then-chromized steel workpiece can cause undesirable consequences during technological applications because soft layer (i.e., carbon-depleted region)

Fig. 5.8 Schematic diagram illustrating the mechanism for formation of the surface layers during carburizing and then chromizing of mild steel specimen

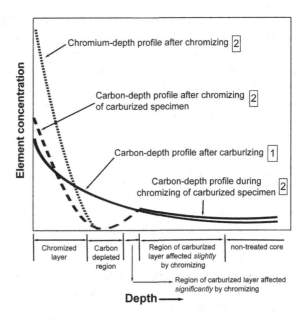

Fig. 5.9 Schematic diagram illustrating the mechanism for formation of the surface layers during chromizing and then carburizing of mild steel specimen

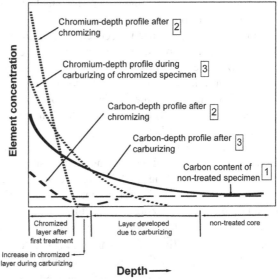

is embedded between much harder surface layer and untreated core. Under various types of loading conditions, carbon-depleted region can undergo a plastic deformation at much lower stress than that of required for the deformation of surface and/or core. This type of situation is favorable for early failure of the workpiece due to the delamination of hard surface-layer from the substrate and, also, due to the formation of subsurface cracks within the soft carbon-depleted region. In case of the carburized workpiece, it can undergo decarburization at elevated temperature and, hence loses its surface hardness. The chromizing-then-carburizing treatment seems to have advantages over the single carburizing process appealing in this regard because this treatment eliminates the carbon-depleted region and the surface region can be relatively stable even at elevated temperatures due to the formation chromium-carbides. However, surface hardness of the duplex treated specimens is lower than the chromized specimen (but it is more than carburized specimen: cf. Table 5.1).

5.5 Conclusions

- Upon chromizing mild-steel, chromium-carbide layer formation has been observed at the surface. These chromium carbides are $Cr_{23}C_6$, Cr_2C, and Cr_3C_2. Absence of pearlite colonies in the region between chromium-carbide layer and nontreated core confirmed the presence of carbon-depleted region, which is responsible for undesirable hardness-depth profile across the cross-section of specimen, i.e., abrupt decrease and then, again increase in hardness with increasing depth from the surface.
- In "duplex surface treatment" of mild-steel, the sequence of the treatment has an important role on the microstructure and, hence, hardness-depth profile. Similar to the chromized steel, carburized-then-chromized specimen showed undesirable occurrence of carbon-depleted region. However, in case of the chromized-then-carburized specimen, carbon-depleted region, which is formed during chromizing, is filled with carbon during carburizing. Therefore, the cross-section of chromized-then-carburized specimen consists of chromium-carbide layer followed by pearlite colonies. Such microstructure causes the continuous decrease in the hardness with depth.
- Significant improvement in the surface hardness has been observed after surface treatments. Compared with the nontreated specimens, the improvement is about 82 % after carburizing, 370 % after chromizing, 205 % after carburizing-then-chromizing and 141 % after chromizing-then-carburizing.
- Square of the carburizing and chromizing depth is approximately linear against the treatment time. For mild steels carburized at 1,000 °C (carburizing depth) $= 455.2\sqrt{\text{time}}$, where carburizing depth is expressed in μm and time in h. Similarly, for mild steels chromized at 1,000 °C (chromizing depth) $= 3.1\sqrt{\text{time}}$, where chromizing depth is expressed in μm and time in hour.

References

1. ASM Handbook 5 (1994) Surface engineering. ASM International, Materials Park, OH, USA
2. Budinski KG (1998) Surface engineering for wear resistance. Prentice Hall, Englewood Cliffs, New Jersey

3. Lee J, Duh J, Tsai S (2002) Surf Coat Technol 153:59
4. Wang ZB, Lu J, Lu K (2005) Acta Mater 53:2081
5. Zhou Y, Chen H, Zhang H, Wang Y (2008) Trans Nonferrous Met Soc China 18:598
6. Peng X, Yan J, Dong Z, Xu C, Wang F (2010) Corros Sci 52:1863
7. Peng X, Yan J, Xu C, Wang F (2008) Metallurg Mater Trans 39A:119
8. Bell T, Dong H, Sun Y (1998) Tribol Int 31:127
9. Tsujikawaa M, Yoshidab D, Yamauchic N, Uedac N, Sonec T, Tanaka S (2005) Surf Coat Technol 200:507
10. Fabijanic DM, Kelly GL, Long J, Hodgson PD (2005) Mater Forum 29:77
11. Kadlec J, Dvorak M (2008) Strength Mater 40:118
12. Dong H (2010) Surface engineering of light alloys: aluminum, magnesium and titanium alloys. CRC Press, Australia
13. Zhang X, Fan A, Zhu R, Ma Y, Tang B (2011) IEEE Trans Plasma Sci 39:1598
14. Kumar V, Hosmani SS (2011) J Metallurg Mater Sci 53:393
15. Vinay K (2010) Thermochemical heat treatment of pure iron: microstructure and kinetics. M.Tech. Dissertation, National Institute of Technology–Karnataka, Surathkal, Mangalore, India
16. ASM Specialty Handbook (1995) Tool materials. ASM International, Materials Park, OH, USA

Chapter 6
Characterization of Surface-Treated Materials

Abstract In engineering applications, surface coatings or treatments are required to increase the lives of machines or components exposed to abrasion or erosion and also to increase the anticorrosion, fatigue, and corrosion fatigue-resistance performance of the machining tools and dies. The thicknesses of coatings or surface-treated layers may vary from less than 100 μm to 100 mm. Hard, wear-resistant, and low friction coatings on the substrate or components can be obtained by using different processes such as electrochemical or electroless methods, thermochemical, spray technologies, physically vapor deposition (PVD), and chemically vapor deposition (CVD) techniques. The hard coatings mainly consist of oxides, nitrides, carbides, borides, or carbon. Surface treatment methods are also popular to enhance various properties of the engineering components. These surface layers or coatings are analyzed using X-ray diffraction (XRD), X-ray photoelectron spectroscopy (XPS), scanning electron microscope (SEM), transmission electron microscope (TEM), scratch test, microindentation, nanoindentation, pin-on-disc tribometer, etc. In this chapter, emphasis is given on a brief discussion of the characterization techniques used for the measurement of crystal structure, grain size, compositions, hardness, elastic modulus, thickness, surface composition, and morphology of the surface layers or coatings or films.

Keywords XRD · XPS · SEM · TEM · Nanoindentation

6.1 X-Ray Diffraction

X-ray diffraction (XRD) is an important characterization tool to determine crystal structures, lattice constants, and percentage crystallinity of thin films/coatings. It is also used to find the preferred orientation of polycrystals, defects, stresses, etc., present in coatings. In XRD, a collimated beam of X-rays with a wavelength typically of the order of 0.7–2 Å is incident on a specimen. These wavelengths are comparable to the crystal lattice spacing and strongly diffract by a crystal.

S. S. Hosmani et al., *An Introduction to Surface Alloying of Metals*,
SpringerBriefs in Manufacturing and Surface Engineering,
DOI: 10.1007/978-81-322-1889-0_6, © The Author(s) 2014

The X-ray beam is diffracted by the crystalline phases of the specimen according to Bragg's law as shown in Eq. (6.1),

$$n \cdot \lambda = 2d \sin \theta$$
$$\text{when } n = 1, \quad \lambda = 2d \sin \theta \tag{6.1}$$

where d is the spacing between atomic planes of the crystalline phase, λ is the X-ray wavelength, and n is the order of diffraction. For the first order of diffraction n is equal to one. The intensity of the diffracted X-rays is measured as a function of the diffraction angle 2θ. The resultant diffraction pattern (also known as XRD pattern) is used to identify the crystal structure, crystallite size, and stresses (if any) in the specimen under investigation. The XRD is widely used in materials characterization because it is nondestructive and does not require sample preparation. The resultant XRD pattern indicates a series of characteristic peaks of varying intensity that satisfies Bragg's law. Each pattern is a fingerprint for material identification. XRD gives bulk analysis and the depth of measurement is typically between 10 and 100 μm.

To identify a crystal structure, X-ray spectra are recorded for rotations around three mutually perpendicular planes of the crystal. This provides comprehensive information about the various crystallographic planes of the lattice, and then these data are converted using Fourier transformation to find out the positions of the atoms in the unit cell. This procedure can identify one of the 230 crystallographic space groups corresponding to the structure and lattice constants a, b, c of the unit cell. In diffraction from a powder specimen, fine grains of any crystalline phase will diffract at different Bragg angles in a sequence of diffraction cones, which are intercepted in turn by the rotating detector. If the crystalline grains are not randomly oriented in space, but possess some preferred orientation (crystalline texture), the diffraction pattern will show intensity anomalies, i.e., their intensities will be different from those of standard peaks. In case of thin films or coating, some of the reflection or diffraction peaks may be missed (or have less intensity) in the XRD patterns compared to standard XRD patterns. This indicates that the deposited films must have a preferred orientation. The Joint Committee of Powder Diffraction Standards (JCPDS), now called the International Centre for Diffraction Data (ICDD), is a database of experimentally observed and calculated diffraction spectra and lists both d-spacings and relative intensities. These diffraction data can be compared with the measured spectrum in order to identify the phases present in an unknown sample. JCPDS has compiled an index of spectra for more than 100,000 elements and compounds. One needs to match their unknown spectra against these standards for identification. The absence of discrete sharp peaks within a spectrum (i.e., broad hump) indicates the amorphous or glassy state of the materials. When the sample is not in powder form, the specimen is rotated to reduce preferred orientation, and the 2θ angle is scanned. For a known X-ray wavelength and experimental θ obtained from the XRD pattern, *d-spacing of the particular set of plane for cubic crystals* can be evaluated using Eq. (6.2),

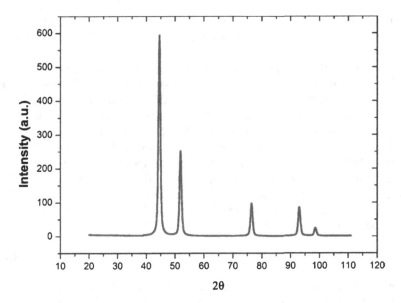

Fig. 6.1 X-ray diffraction pattern from Ni powder [1]

$$d = \frac{a}{\sqrt{h^2 + k^2 + l^2}} \qquad (6.2)$$

where a is the lattice parameter and $h\ k\ l$ are the miller indices of the planes. To identify the unknown sample, we need to first generate a table of the d-spacings calculated from the strong reflections by using Bragg's law and the known wavelength of the X-ray (for example, for CuKα radiation, it is 0.154 nm).These d-spacings are fed into the computerized JCPDS database, and the output identifies the possible phases that best match the experimentally observed d-spacings and their relative intensities. Then, we must extract the data for each of the selected options from the JCPDS database to compare the standard with the measured values of d-spacing and intensity. The typical XRD pattern (intensity versus 2θ) of nickel powder is shown in Fig. 6.1.

To identify the crystal structure of a cubic material, the ratios of $\sin^2\theta$ are calculated. The values of this ratio for BCC, FCC, and DC are 2:4:6…; 3:4:8:11: …; and 3:8:11:16: …, respectively. In other words, for BCC lattice, the planes that produce diffraction peaks are those for which $h + k + l = n$, an even integer, and for FCC lattice the observed diffraction lines either have all odd integers or all even integers [2, 3]. The lattice parameter for this material can be calculated using Eqs. (6.1) and (6.2).

From the XRD pattern of any material, we can also determine the exact lattice parameter. Any variations in lattice parameter indicate either the presence of residual stress (in case of films or coating) or alloying elements' addition in solid solution. By careful calibration of the spectrometer, and assuming a linear dependence of the lattice parameter on composition (i.e., Vegard's law), we can also determine the concentration of the alloy from an exact determination of the lattice parameter [3].

It is also used to characterize homogeneous and inhomogeneous strains. Homogeneous or uniform elastic strain shifts the positions of diffraction peaks. From the shift in peak positions, one can calculate the change in d-spacing. The d-spacing is changed due to the change in lattice constants under an applied load (or strain). In contrast to this, inhomogeneous strains broaden the diffraction peaks. The peak broadening increases with increasing diffraction angle (i.e., $\sin \theta$). It is also caused by the finite size of crystallites, but here the broadening is independent of $\sin \theta$. When both crystallite size and inhomogeneous strain contribute to the peak width; these can be separately determined by careful analysis of peak shapes. If there is no inhomogeneous strain, the grain size of nanocrystalline coating or crystallites size (D) can be determined using Scherrer's formula as shown in Eq. (6.3),

$$D = K \cdot \lambda / (B \cdot \cos \theta) \tag{6.3}$$

where B is the full width at half maximum (FWHM) of a diffraction peak and K is the shape factor. For cubic materials, its value is about 0.9. However, in case of nanosized particles, the presence of twinned structures may cause results different from the actual particle (or grain) sizes. Moreover, peak broadening may also occur due to instrument error, lattice strain (as discussed above), and stacking faults/dislocations. It is to be noted that the Scherrer formula is limited for crystallites of size less than 100 nm.

Instrumental error (or broadening) occurs due to slit widths, sample size, penetration into the sample, imperfect focusing, and non-monochromaticity of the X-ray beam. To correct for instrumental broadening, standard peaks are obtained using a silicon standard in which the crystal size is large enough to eliminate all crystal size broadening. These standards peaks (111) at $2\theta = 28.47°$ and (311) at $2\theta = 56.17°$ are run under instrumental conditions identical with those for the test samples. By assuming Gaussian shapes for the diffraction peaks, the corrected FWHM without instrumental broadening, B_c, can be obtained using Eq. (6.4),

$$B_c = B_h^2 - B_s^2 \tag{6.4}$$

where B_h is the broadening from samples containing both the desired broadening and the instrumental broadening and B_s is the instrumental broadening obtained from the standard. Once the value of B_c is known, the crystal size can be easily determined using Eq. (6.3) by replacing B with B_c. Even after making the correction, the average grain size values determined by XRD might be somewhat larger than the value determined by transmission electron microscope (TEM) histogram. Thus, XRD can estimate average grain sizes, but a TEM is needed to determine the actual distribution of grain sizes.

Figure 6.2 shows the powder XRD pattern of a series of indium phosphide (InP) nanoparticles of different sizes. It can be clearly seen from Fig. 6.2 that the

Fig. 6.2 Powder XRD
pattern of InP nanocrystals of
different sizes. The stick
spectrum shows the bulk
reflections with relative
intensities [4]

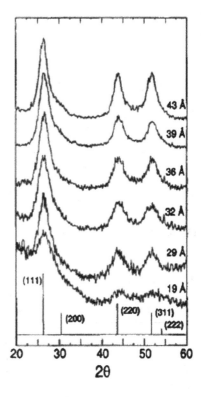

intensity and sharpness of the reflection peaks decrease with decreasing crystallite
size (i.e., from 43 to 19 Å) of InP nanoparticles [4].

XRD patterns of ZnSe thin films grown at 500 and 550 °C on glass substrate is
shown in Fig. 6.3, which indicates the presence of a mixture of both cubic ZnSe
(ICDD 00-037-1463) and hexagonal ZnSe (ICDD 01-080-0008). The intensities of
some of the diffraction peaks from thin films are different from those of bulk. This
is due to the texture or orientation of the grains in the film [5].

Figure 6.4a shows XRD patterns of Mg powder and its coatings. The relative
intensity of crystal plane (0002) in XRD patterns of the coatings is significantly
enhanced compared to that of the powder. This is probably due to the texture
caused by the plastic deformation of the particles during impact of the coating
process, i.e., the grains are turned to some direction, and thus the plane (0002) also
turns to this direction. The occurrence of the texture confirms the plastic deformation of the particles [6].

The effect of alloying elements or doping on the positions of reflections can be
studied using XRD. Figure 6.4b shows the XRD patterns of ZnO and Al-doped
ZnO nanoparticles. In all the doped nanoparticles, the peak positions of reflections
do not change indicating that all aluminum ions are incorporated into the crystal
lattice of bulk ZnO to substitute for zinc ions. However, the width of diffraction
peaks increases with the increase in Al doping content. From Scherrer's formula

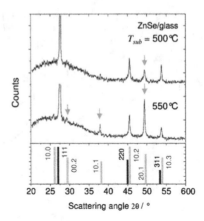

Fig. 6.3 XRD patterns of the ZnSe thin films grown at 500 and 550 °C on glass [5]

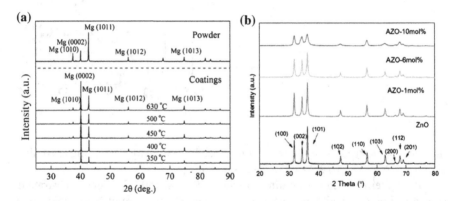

Fig. 6.4 Shows XRD patterns of **a** Mg powder and its coatings [5], and **b** ZnO and Al-doped ZnO (AZO) nanoparticles [7]

Eq. (6.3), we know that crystallite size is inversely proportional to the FWHM of a diffraction peak. This indicates that the crystalline size of ZnO nanoparticles decreases with increasing content of Al ions. The average grain size can be estimated using Eq. (6.3) provided the films or coatings have average grain sizes smaller than 100 nm and they are free from any strains [7]. In case of larger grains, average grain sizes are determined by measuring at least 300 grains in the scanning electron microscope (SEM) micrographs using the method of linear intercept. However, the use of optical microscope is cost-effective and fast compared to SEM.

Figure 6.5a shows XRD patterns of coatings A1, A2, A3, and A4 obtained under four different conditions of (Ti + N_2) reactive deposition during sputtering process. The N_2 content in the sputtering gas increases from A1 to A4, i.e., the nitrogen atomic contents increased from 0 to 43.3 at.%. Increasing the nitrogen partial

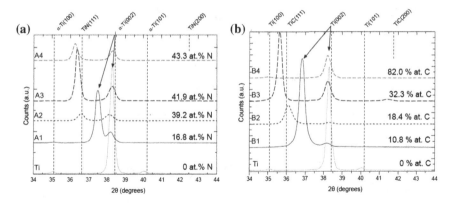

Fig. 6.5 XRD patterns of **a** nitrogen doped Ti and **b** carbon-doped Ti. The N_2 content increases from A1 to A4 and similarly, carbon content increases from B1 to B4 [8]

pressure leads to greater nitrogen contents in the films. The sputtering process is typically a nonequilibrium process and supersaturated metastable Ti (i.e., α-Ti) with 16.8 at.%N (i.e., >equilibrium solubility of 10 at.%.) in solution is obtained. The α-Ti (002) peak is shifted to the left due to nitrogen supersaturation at the octahedral sites of the hexagonal unit cell. This indicates that the d-spacing of the Ti-unit cell is increased which in turn increases the lattice constant. On further increasing the nitrogen content, the TiN is formed and showed NaCl type unit cell for the A2, A3, and A4 conditions. The plane (111) orientation turns out to be very strong for the A3 film. Similarly, the increase in the carbon partial pressure (Ti + CH_4) leads to greater carbon contents in the film as shown in Fig. 6.5b. The maximum equilibrium solubility of carbon in (α-Ti) at 300 °C is 0.07 at.%. The carbon content increases from B1 to B4. There is formation of supersaturated hexagonal Ti (α-Ti) containing 10.8 at.% carbon. In case of B2 and B3 films, 18.4 at.% and 32.3 at.% carbon are dissolved in the NaCl type TiC phase, respectively. A slight shift of the (111) plane to lower 2θ values can be attributed to changes in chemical composition and tensile stresses in the TiC films. The film B3 showed a strong texture in the (111) plane. However, when the carbon content is 82 at.%, diffraction peaks from α-Ti were detected, but not from TiC phase. This is probably due to a low TiC volume fraction and crystallite size, i.e., of the order of nanometers [8].

Figure 6.6 shows the XRD patterns of the as-deposited and annealed tin-doped In_2O_3 (ITO) thin films. Films were prepared by an electron beam evaporation system in the presence of oxygen and annealed at annealing temperatures of 200 and 300 °C. The inset of Fig. 6.6 shows a single broad background peak centered on $2\theta = 25.8°$. This feature indicates either the presence of an amorphous phase and/or a nanocrystalline structure. As annealing temperature is increased from 200 to 300 °C, the number and intensity of peaks are increased due to the increase of grain size [9].

It is to be noted that XRD provides the collective information about the particle/grains from a sizable amount of powder/coating. The film thickness of epitaxial and

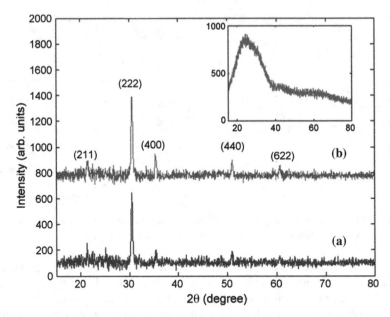

Fig. 6.6 XRD patterns of tin-doped In_2O_3 samples after annealing at **a** 200 and **b** 300 °C in air for 1 h. Inset is the XRD pattern of the as-deposited ITO thin films [9]

highly textured thin films can also be determined using XRD method. Compared to electron diffraction, one of the disadvantages of the XRD is the low intensity of diffracted X-rays, particularly for the materials with low atomic number. XRD is more sensitive to materials with high atomic number. The typical intensities for electron diffraction are about eight orders of magnitude higher than that of XRD. Hence, XRD generally requires large volume or area of the specimens.

6.2 X-Ray Photoelectron Spectroscopy

X-ray photoelectron spectroscopy (XPS) is the most useful analysis tool to obtain chemical information about different elements at sample surfaces. In XPS, the sample is placed in a high vacuum environment and the electron beam of high-energy radiation (e.g., X-ray or UV) irradiates the sample and produces photo-electrons. The energy of the ejected photoelectrons is a function of its binding energy (i.e., the energy required to remove the electron from an atom) and is characteristic of the element from which it was emitted. The binding energy (E_B) of the photoelectrons can be determined using Eq. (6.5),

$$E_B = h\upsilon - E_K - W \qquad (6.5)$$

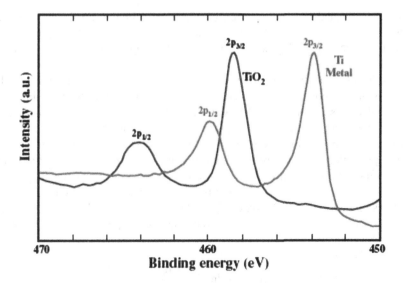

Fig. 6.7 XPS spectra of Ti and TiO$_2$ [1]

where E_B is the binding energy of the electron from a particular energy level, hυ is the incident X-ray photon (or electron) energy, E_K is the kinetic energy of the ejected photoelectron, and W is the spectrometer work function. From the binding energy we can obtain some important information, such as relative quantity and type of each element, the chemical state of the elements present, depth distribution, etc., about the samples under investigation. XPS can characterize the surface compositions to a depth of 0.5 ~ 1 nm with a spatial resolution of 0.2 mm, and a sensitivity of 0.3 %. By employing a monochromatic source of known energy like Mg K$_\alpha$ (1254 eV) or Al K$_\alpha$ (1487 eV) X-rays and measuring the energies of the emitted photoelectrons, one can find the binding energies of the electrons. The binding energies of the electrons depend on the environment of the atoms and its oxidation state. By comparison with the standard binding energies, conclusion can be drawn on the bonding between atoms. Since photoelectric effect occurs from the surface, the technique is helpful to analyze the chemical composition on the surface. For example, XPS has proved the formation of SiC in carbon nanotubes (CNTs) coated with Si by the CVD process at 1,000 °C. Binding energies corresponding to Si-C and Si-O bonds are different. The binding energies of the electrons in the outermost shell of an atom are sensitive to the chemical state of the atom, the strength, and nature of the chemical bonds. Hence, the XPS can be used to get the oxidation states of metals. Figure 6.7 shows the XPS spectra of Ti and its oxide (i.e., TiO$_2$) and also indicates different 2p states in a titanium atom. The formation growth of TiO$_2$ on titanium metal is accompanied by 2p peak shifts of 4–5 eV. The higher binding energy in the oxide results in a large peak shift.

XPS is also an important tool for characterizing the wear surfaces of metal-carbon nanotube (CNT) composites. XPS analysis of the wear track of Al-CNT composite revealed that higher Al_4C_3 content with higher CNT reinforcement leads to poor wear resistance. The identification of intermediate phases and compounds forming during wear of metal-CNT composites can also be studied with XPS.

XPS is also widely used to measure the thickness of a surface film. The thickness of a homogeneous film is calculated from the ratio of the peak intensity of photoelectrons emitted from the overlayer thin film and from the substrate. The inelastic mean free path for the peak from the substrate and that from the overlayer thin film is assumed to be the same. The kinetic energy of elemental or metallic state of photoelectron peak from the substrate is almost the same as that of oxide state photoelectron peak from the overlayer thin film. The elemental Si 2p peak and oxide Si 2p peak can be extracted by curve fitting the measured Si 2p peak. The thickness (t) of the silicon oxide (SiO_2) layer on the Si can be calculated using Eq. (6.5). Here, the presence of suboxides on the interface between oxide layer and substrate is ignored.

$$t = L_{SiO_2} \cos \theta \ln \left[1 + \frac{(I_{SiO_2}/R_{SiO_2})}{I_{Si}} \right] \qquad (6.6)$$

Here, L_{SiO2} is the attenuation length of Si_{2p} photoelectrons in an SiO_2 layer. The values of L_{SiO2} for Mg Kα X-ray and Al Kα X-ray are 2.964 and 3.448 nm, respectively, θ is the emission angle of the photoelectron with respect to the surface normal, I_{SiO2} is the peak intensity of oxide state Si_{2p}, I_{Si} is the peak intensity of metal state Si_{2p}, and R_{SiO2} is the intensity ratio of the bulk Si oxide to the bulk Si, which is 0.9329 [10].

6.3 Scanning Electron Microscopy

Scanning electron microscopy (SEM) is the most widely used technique in characterization of bulk and/or thin films. SEM provides high-resolution images of surfaces of samples using electrons instead of light waves in optical (or Metallurgical) microscopes. The advantages of SEM over optical microscope include higher magnification and greater depth of field ($\sim$microns). The maximum magnifications of up to 1,000,000× with a resolution of better than 5 nm can be achieved. By contrast, the maximum useful magnification of an optical microscope is about 1,000× with its resolution not better than 1 µm. In SEM, the incident electrons (from an electron gun) typically have energies of 2–40 keV.

SEM images the sample surface by scanning it with a high-energy beam of electrons in a raster scan pattern. The electrons interact with the atoms that make up the sample producing signals that contain information about the sample's surface topography and composition. When electron beam hits the sample, secondary or back-scattered electrons are produced, which are collected by a secondary detector or a backscatter detector, respectively. These electrons are

Fig. 6.8 Schematic representation of possible interactions between the primary electron beam and **a** the nucleus of surface atoms and **b** electrons of surface atoms comprising the sample. For simplicity, only K, L, and M shells (with no L and M sub shells) are shown [11]

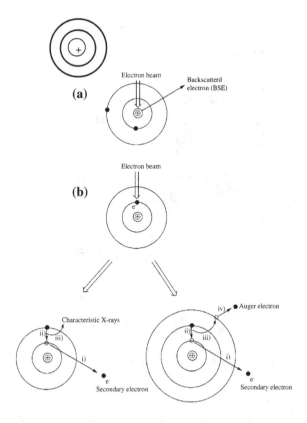

converted to a voltage, and amplified. The amplified voltage is applied to the grid of the cathode ray tube (CRT) and causes the intensity of the spot of light to change. The image consists of thousands of spots of varying intensity on the screen of a CRT that correspond to the topography of the sample. The secondary electron signals result from interactions of the electron beam with atoms at or near the surface of the sample. Due to the narrow electron beam, SEM micrographs have a large depth of field yielding a characteristic three-dimensional appearance useful for understanding the surface structure of a sample. A wide range of magnifications is possible, from about $10\times$ to more than $1{,}000{,}000\times$, about $1{,}000\times$ the magnification limit of the best optical microscopes.

SEM can provide both structural information as well as quantitative analysis from crystalline samples. The interaction among high-energy electrons and sample atoms results in a variety of emissions that yield important information about the surface morphology (Fig. 6.8) and elemental composition of the sample. If the electron beam interacts with the nucleus of surface atoms, the electrons are elastically scattered and the trajectory of the electron changes with no change in the kinetic energy. Such type of interaction is known as *backscattering*. A beam of back-scattered electrons (BSE) is reflected from the sample by elastic scattering. The number of BSE produced from a given atom is proportional to the atomic

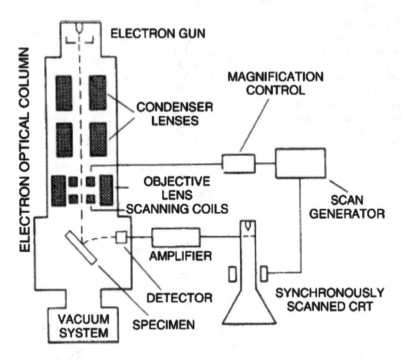

Fig. 6.9 Schematic diagram showing the basic components of the SEM [12]

number. That is, materials composed of heavy atoms will backscatter more electrons, resulting in brighter gray tones in the image relative to less dense materials. Hence, BSE produce an image that is related to material composition, providing both spatial and chemical information. BSE are often used in analytical SEM along with the spectra made from the characteristic X-rays. As the intensity of the BSE signal is strongly related to the atomic number (Z) of the specimen, BSE images can provide information about the distribution of different elements in the sample. Characteristic X-rays are emitted when the electron beam removes an inner shell electron from the sample, causing a higher energy electron to fill the shell and release energy. These characteristic X-rays are used to identify the composition of the sample.

When the primary electron beam interacts inelastically with electrons from surface atoms, the collision displaces core electrons from filled shells. This results in atoms with an energetic excited state, which have missing inner shell electron. These are known as secondary electrons having low energy compared to BSE.

Figure 6.9 shows a schematic diagram of the SEM consisting of basic components. A fine probe of electron beam is scanned across a selected area of the specimen surface in a raster by scan coils. The electrons penetrate the specimen in a teardrop-shaped volume. The penetration depth of the interaction volume increases with increasing energy of the electron beam. The interaction of the electron beam with the specimen produces secondary, backscattered and Auger electrons, X-rays, etc., which are collected by various detectors in the specimen chamber. The signal from each detector can be fed to a monitor, which is rastered in synchronization with the electron beam. The magnification of the image is determined by the ratio of the side length of the monitor display to the side length of the raster on the specimen. The best resolution for secondary electron images achieved in field emission gun SEM (FEG-SEM) is about 1 nm, but it is typically 5 nm or poorer than this for SEM consisting of LaB_6 and W filaments. SEM can generate high quality images with less sample preparation.

SEM can also be used to assess the surface morphology of amorphous carbon films. However, it does not convey much detail about the film structure as amorphous carbon films usually display a smooth morphology. Moreover, its electrical conductance is often low and hence, surface charging can occur and distort the resulting SEM image. In general, SEM is used to determine nano-structured compounds or particles formed on the amorphous carbon film surface. It is also used to observe the cross-section of a film to determine the film thickness, film growth rate, and film growth mechanism. It is an excellent tool to observe the nature of transfer film on the counter surface, friction track, and analyze the debris after a wear test for a given sliding distance and load [11–13].

Users need to remember that the samples must be of appropriate size to fit in the specimen chamber. The samples are generally mounted rigidly on a specimen holder or metallic stub. SEM can examine any part of a sample having length up to 15 cm, and can tilt a specimen of this size to 45°. For conventional imaging, specimens must be electrically conductive, at least at the surface, and electrically grounded to prevent the accumulation of electrostatic charge at the surface. Metal objects require little special preparation for SEM except for cleaning and mounting on a specimen stub.

Insulating specimens like polymers and ceramics tend to charge when scanned by the electron beam. They are therefore usually coated with a very thin ($\sim$5–10 nm thick) coating of electrically conductive material like gold, gold/ palladium alloy, platinum, and carbon. They are generally deposited on the sample either by low vacuum sputtering coater or high vacuum evaporation unit. Coating prevents the accumulation of static electric charge on the specimen during electron-specimen interaction. However, insulating specimens may be imaged without coating using "Environmental SEM" (E-SEM) or FEG-SEM operated at low voltage. E-SEM places the specimen in a relatively high pressure chamber where the working distance is short and the electron optical column is differentially pumped to keep the vacuum adequately low at the electron gun. The high pressure region around the sample in the E-SEM neutralizes charge and provides an amplification of the secondary electron signal. FEG-SEM can be operated at low voltage because the FEG is capable of producing high primary electron brightness even at low accelerating potentials. To examine the thickness of thin coating, the specimen may be embedded in a polymer resin with further polishing to a mirror-like finish.

SEM is widely employed to examine fracture modes and fracture mechanisms. The fractured surface is cut to a suitable size, cleaned of any organic residues, and mounted on a specimen holder for viewing BSE image in the SEM. Heavy elements backscatter electrons more strongly than light elements, and thus appear brighter in the image. BSE are used to detect contrast between areas with different chemical compositions.

Characteristic X-rays are generated by atoms when the incident high-energy electrons knock out inner shell electrons and outer shell electrons move into the empty orbit of the specimen. There is now a progression of electron jumps from higher to lower energy states (e.g., from L to K, then M to L, etc.) until all the electron states are refilled. At each stage, X-rays are emitted to conserve energy. Measurement of the energies (or wavelengths) of these X-rays gives information about the chemical composition of the specimen. Characteristic X-rays are emitted from the entire specimen–beam interaction volume, so resolution is no better than 1 μm or so. The X-rays are detected by either an energy dispersive X-ray spectrometer (EDS or also known as EDX) or a wavelength dispersive spectrometer (WDS). The EDS is the more common attachment to SEM or TEM as it provides rapid qualitative analysis of the specimen. Semi-quantitative analysis can be obtained with care, but accurate microanalysis requires a WDS and use of standards to allow for X-ray absorption in the specimen and fluorescence (the excitation of lower energy X-rays by higher energy ones). An "Electron Microprobe" is a SEM fitted with WDS and analysis software and is dedicated to the analysis of chemical analysis. WDS uses the diffraction patterns created by electron beam–specimen interaction as its raw data. It has a finer spectral resolution than EDS. In WDS, only one element can be analyzed at a time, while EDS gathers a spectrum of all elements of a sample at a time [13, 14].

Figure 6.10 shows a SEM micrograph from a polished and thermally etched alumina specimen, together with an EDS spectrum obtained from the entire region shown in the micrograph. The alumina specimen was doped with a small amount of magnesium, silicon, and calcium. The EDS can easily detect the main aluminum and oxygen peaks. The absence of Mg, Si, and Ca peaks in the EDS spectrum was expected, since the total concentration of these dopants and impurities in a large sample volume is below the detection limit of the EDS system. We cannot see any of the additives within the grains because the solubility limit of these cations is well below the detection limit of EDS. However, an EDS spectrum, taken with the probe positioned at a single point located at a grain boundary, clearly shows the presence of both Mg and Si. It implies that the alumina grains are saturated with these impurities, although the amount of Ca at this grain boundary is still below the detection limit [1].

6.4 Transmission Electron Microscopy

In TEM, the electrons are generated from an electron gun, interact with specimen and are scattered. The scattered electrons are focused using electron optic lenses to finally form images. The imaging modes can be controlled by the use of an aperture.

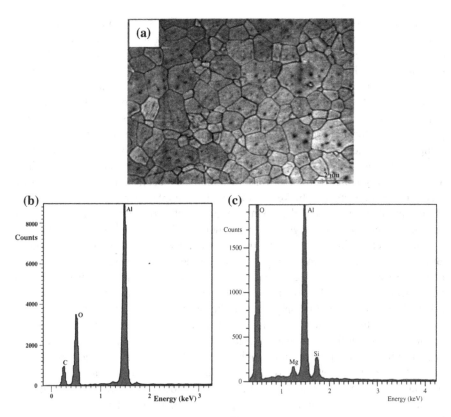

Fig. 6.10 **a** Secondary electron image from thermally etched alumina **b** EDS spectrum from the region shown in (**a**), and **c** EDS point spectrum recorded from a grain boundary in the alumina specimen [1]

If most of the scattered electrons are allowed through, we get Bright Field image. If specific scattered beams are selected, the image is known as a Dark Field image. In addition, a TEM can also be used for chemical analysis. Electrons scattered from the sample are collected on a CRT to form the image. Selected area diffraction (SAD) or convergent beam electron diffraction (CBED) is used to characterize the crystalline nature of samples from areas as small as microns (SAD) or tens of nanometers (CBED) via electron diffraction patterns. Analytical TEM can provide elemental analysis, maps, and line scans using auxiliary detectors or attachments. These detectors allow them to analyze X-rays or energy spectra of the secondary emission electrons and thus, act as a powerful tool for the study of qualitative and quantitative chemical analysis. Some of the attachments are EDS, WDS, electron back-scattered diffraction (EBSD), XPS, electron energy loss spectroscopy (EELS), and Auger electron spectroscopy (AES). The EELS is a better technique than EDS for identification of phases containing light elements like carbon, nitrogen, oxygen, etc., at high spatial resolution of ~ 1 nm.

TEM consists of an electron emission source. The source may be a tungsten filament or lanthanum hexaboride (LaB_6). Applying a high voltage source typically in the range of 100–300 kV can give sufficient electron current either by thermionic or field emission into the vacuum. The upper lenses of the TEM allow for the formation of the electron probe to the desired size and location for later interaction with the sample. Typically, a TEM consists of condenser lenses, objective lenses, and projector lenses. The condenser lenses are responsible for primary beam formation, while the objective lenses focus the beam down onto the sample. The projector lenses are used to expand the beam onto the phosphor screen or photographic film. Both bright field and dark field images can be obtained in a TEM. The bright field image is formed directly by occlusion and absorption of electrons in the sample. Thicker regions of the sample, or regions with a higher atomic number will appear dark, while regions with no sample in the beam path will appear bright. The image is a two-dimensional projection of the sample.

Samples can exhibit diffraction contrast, whereby the electron beam undergoes Bragg scattering, which in the case of a crystalline sample disperses electrons into discrete locations in the back focal plane. By the placement of apertures in the back focal plane, i.e., the objective aperture, the desired Bragg reflections can be selected (or excluded), thus only parts of the sample that are causing the electrons to scatter to the selected reflections will end up projected onto the imaging apparatus. If the reflections that are selected do not include the unscattered beam (which will appear up at the focal point of the lens), then the image will appear dark wherever no sample scattering to the selected peak is present, as such a region without a specimen will appear dark. This is known as a "dark-field" image. Modern TEM is also equipped with a specimen holder that can allow the user to tilt the specimen to a range of angles. This may help to obtain specific diffraction conditions, and apertures placed above the specimen allow the user to select diffracted electrons. In addition to the microstructure on a fine scale, TEM also identify lattice defects in crystals [14].

6.5 Auger Electron Spectroscopy

Auger electron spectroscopy (AES) can be utilized to characterize the surface impurities to a depth of 0.5 ~ 1 nm with a spatial resolution of 0.2 μm, and a sensitivity of 0.3 %. The AES is a useful instrument where electrons emerge or escape from the top 2 nm of the sample. AES is truly a surface science tool. It is capable of detecting all elements above Li. The Auger electron transition energies and associated spectra for each pure element are compiled into an Auger Handbook. Cross reference of an unknown spectrum against the handbook standards makes elemental analysis fairly straightforward. The relative sensitivity of each element's Auger yield is also compiled and therefore semi-quantitative analysis is possible. It is also capable of elemental detection to approximately 1 % in solution. Perhaps one of the most versatile ways of using this technique is in conjunction with ion sputtering. Although the sample is damaged by the sputtering process, one obtains an elemental map normal to the sample surface. This is commonly referred to as depth profiling, which can be obtained with high purity Argon gas [15].

6.6 Coating Adhesion

Coating adhesion is mostly measured using the tape method, pull-off test, shock wave loading method, and scratch test. Of these methods, scratch testing is a reliable test. A scratch tester has a smoothly rounded chrome-steel stylus with a tungsten carbide or Rockwell C diamond tip (in the form of a 120° cone with a hemispherical tip of 0.2 mm radius). The Vickers microhardness tester can also be used as scratch tester by mounting a stylus with a diamond tip. The diamond indenter is lubricated by a fine watch oil during the scratching operation. After the scratch is made, the specimen is cleaned of oil. This stylus is drawn across the coating surface, and a vertical load is applied to the point and is increased in step until the coating is completely removed [16]. The minimum or critical load at which the coating flakes (adhesive failure) or chips (cohesive failure) is used as a measure of adhesion. To avoid damage to the diamond, measurement of unknown substances should be started with the smallest force. Adhesion is measured on the basis of critical load values. The width of the scratch is read in microns by means of a filar micrometer eyepiece. The scale is derived by using the reciprocal of the cut width in microns squared, multiplied by 10,000;

$$K = 10,000/W^2 \qquad (6.7)$$

where K is the microcharacter scale and W is the width of cut in microns.

6.7 Measurement of Mechanical Properties

6.7.1 Microindentation Hardness Tester

In microhardness test, the applied load and the resulting indent size are small relative to bulk tests. The most commonly used microindentation tests are the Vickers and Knoop tests. In 1925, Smith and Sandland of the UK developed the Vickers hardness test. The Vickers hardness test method consists of indenting the test material with a pyramidal diamond indenter having a square base. Its opposite faces have an angle of 136°. The applied loads may vary between 1 and 100 kgf. The dwell time for the applied load is typically 10–15 s. After removal of the applied load, the two diagonal lengths of the indentation left in the material are measured using an optical microscope and their average is taken. In general, the impression of the indentation appears to be square. Vickers hardness number is calculated based on the surface area of the indentation. It is to be noted that the area of the sloping surface of the indentation is calculated using Eq. (6.8). The Vickers hardness is defined as the quotient obtained by dividing the load with area of indentation,

$$HV = [2F \sin(136°/2)]/d^2 = (1.854\,F)/d^2 \qquad (6.8)$$

where F is the Load (in kgf), d is the mean value of the two diagonal lengths (in mm), and HV indicates Vickers hardness. The Vickers hardness (HV) number of 500 measured under a 10 kgf load is reported as 500 HV_{10}. The other obsolete symbols are DPN or VPN. There are many advantages to the Vickers hardness test. It provides extremely accurate readings for testing the softest to the hardest materials. In this, one type of indenter can give hardness for all types of metals and surface treatments. One of the disadvantages of the Vickers machine is its higher cost than the Brinell or Rockwell machines. Vickers hardness number can be reported in SI units (MPa or GPa). To convert HV number into MPa and GPa, one has to multiply this number by 9.807 and 0.009807, respectively. In this, it is assumed that elastic recovery does not occur once the load is removed.

The Knoop indenter is a rhombic-based pyramidal diamond with longitudinal edge angles of 172.5 and 130°. In general, the loads used in the Knoop tester vary from about 0.2 to 4 kg. Smaller loads as low as 1–25 g may also be used in both the testers. In both indentations, the lengths of the diagonals are measured using a medium-power compound microscope after the load is removed. If b is the long diagonal (in mm), the Knoop hardness number HK, can be represented by Eq. (6.9).

$$HK = 14.229\ F/b^2 \qquad (6.9)$$

The Knoop hardness is also expressed in the same manner as the Vickers hardness. A main source of error in the tests is the alignment of the sample surface relative to the indenter. The indenter itself must be properly aligned perpendicular ($\pm 1°$) to the stage plate. The sample surface must be perpendicular to the indenter. Tester holders can be used to align the polished face perpendicular to the indenter. Compared to Vickers indenter, the Knoop indenter gives a longer diagonal length for a given depth of indentation or a given volume of material deformation. In other words, for the same measured diagonal length, the depth and area of the Knoop indentation is only about 15 % of those of the Vickers indentation. Thus, Knoop indenter is preferred for shallow specimens, particularly for brittle materials, such as glass or diamond, in which the tendency for fracture is related to the area of stressed material.

Fig. 6.11 Schematic
illustration of load–
displacement curve [17]

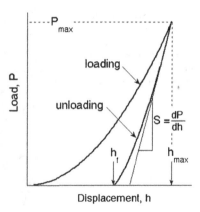

The measurement of hardness and elastic modulus of the materials using
indentation method was reported in 1992. These properties can be determined
directly from the indentation load and displacement curve. SEM has facilitated the
measurement of properties at micrometer to nanometer scales for thin films and
coatings. A typical set of load-displacement data obtained with a Berkovich
indenter is shown in Fig. 6.11, where the parameter P represents load and h is the
displacement relative to the initial undeformed surface. There are three important
parameters such as maximum load (P_{max}), maximum displacement (h_{max}), and
elastic unloading stiffness ($S = dP/dh$). Here, the stiffness is defined as the slope of
the upper portion of the unloading curve during the initial stages of unloading. The
final depth (h_f) is the permanent depth of penetration after the load is removed. As
the indenter is driven into the material, both elastic and plastic deformation occurs,
which results in the formation of a hardness impression to some contact depth (h_c).
When the indenter is withdrawn, the elastic portion of the displacement is
recovered, which facilitates the analysis of hardness and elastic modulus of the
materials.

Once the contact area is determined from the load–displacement curve, the
hardness, H, and effective elastic modulus, E_{eff}, can be calculated using Eqs. (6.10)
and (6.11), respectively [16].

$$H = P_{max}/A = P_{max}/24.5\,h_c^2 \qquad (6.10)$$

where A is the contact area. For a perfectly sharp Berkovich or Vickers indenter, it
can be represented as a function of contact depth, $A = 24.5\,h_c^2$. The P_{max} and h_c can
be directly determined from the load–displacement curve.

$$E_{eff} = \frac{1}{\beta}\frac{\sqrt{\pi}}{2}\frac{s}{\sqrt{A}} \qquad (6.11)$$

where β is a constant which depends on the geometry of the indenter ($\beta = 1.034$
for the Berkovich) [17, 18].

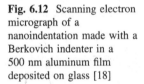

Fig. 6.12 Scanning electron micrograph of a nanoindentation made with a Berkovich indenter in a 500 nm aluminum film deposited on glass [18]

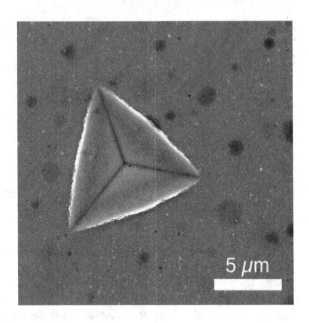

6.7.2 Nanoindentation Hardness Tester

Nanoindentation is a technique routinely used to determine the mechanical properties of thin films or surface layers. It employs high-resolution sensors and actuators to continuously control and monitor the loads and displacements on an indenter as it is driven into and withdrawn from a material. The force as small as a nano-Newton and displacement of about an Angstrom can be precisely measured. Berkovich triangular pyramidal indenter is useful to probe properties at the smallest possible scale. As discussed above, it is made of diamond which has an effective tip radii of the order of 10–100 nm. A scanning electron micrograph of a small nanoindentation made with a Berkovich indenter in a 500 nm aluminum film deposited on glass is shown in Fig. 6.12.

The mechanical properties can be determined from the indentation load–displacement data alone, thereby avoiding the need to image the hardness impression and facilitating property measurement at the submicron scale. The second advantage is that measurements can be made without removing the film or surface layer from its substrate. With special indenters, it is also possible to make indentations in brittle materials with radial cracks extending from the edges of the contact, which allows one to explore fracture behavior at the micron scale and measure fracture toughness in volumes of material a few microns or less in diameter. Sharp indenters are also potentially useful in the study of residual stresses in thin surface regions and layers, although residual stress influences on indentation behavior are usually small and difficult to detect. Spherical indenters are also frequently employed in the measurement of mechanical properties by ultra-low load indentation methods. The primary advantage of the spherical indenter is that indentation contact begins elastically as the load is first applied, but then changes to elastic-plastic at higher loads, thereby allowing one to explore yielding and associated phenomena. Although the point of initial yielding is sometimes difficult to identify experimentally because plasticity commences well below the surface, one can in principle use a spherical indenter to determine the elastic modulus, yield stress, and strain-hardening behavior of a material all in one simple test. Moreover, it is sometimes possible to deduce much of the uniaxial stress–strain curve from spherical indentation data. Another reason for interest in spherical indenters is that even the sharpest Berkovich diamonds are rounded at some small scale. Thus, the indentation behavior of very small Berkovich indentations requires that the influence of the spherical tip geometry on the contact mechanics be considered.

However, it is difficult to obtain precise spherical geometries for diameters less than about 100 microns, particularly for hard materials like diamond from which indenters are made [17, 18].

6.8 Thickness Measurement

The larger sized thicknesses of the coatings may easily be determined using SEM or optical microscopes. However, when the thickness is at nano level, an interferometer is used. An interferometer is an arrangement in which commercial microscopes utilize two-beam or multiple-beam techniques for measuring coating thickness. The wedge or step technique is generally used for thickness measurements of opaque coatings. A step is made in the coating by masking a portion of the substrate during deposition or by removing part of the coating from the substrate. The interference fringes are formed between an optical flat and the surface of the coated specimen using a white light that can be observed with the microscope. If there is a slight angle between the two surfaces of the optical flat and the specimen, the fringe pattern will appear as alternating parallel light and dark bands that are perpendicular to the step on the coatings. Two adjacent fringes are separated by half the light wavelength (λ). The step in the coated specimen surface will displace the fringe pattern, and this displacement is a measure of the difference in the level of the surface of the specimen. The coating thickness is measured as the displacement of the fringe pattern at the step area and this coating thickness is measured using Eq. (6.12),

$$t = (d/s) \, \lambda/2 \tag{6.12}$$

where d is the displacement of the fringe pattern at step area, s is the distance between the two adjacent fringes, and λ is the wavelength of light. The sharpness of the fringes increases significantly if interference occurs between many beams.

This can be achieved if the reflectivity of the two samples at interfaces is very high. While the displacement of the fringes of a double-beam instrument can be estimated to about one-tenth of the fringe spacing, the fringes of a multiple-beam instrument are much more sharply defined, and their displacement can be estimated to perhaps one-hundredth of the fringe spacing (~ 3 nm) [19, 20].

References

1. Brandon D, Kaplan WD (2008) Microstructural characterization of materials, 2nd edn. John Wiley and Sons Ltd, England
2. Cullity BD (1956) Elements of X-ray diffraction. Addison-Wesley Publishing Company, Inc., USA
3. Suryanarayan C, Norton MG (1998) X-ray diffraction: a practical approach. Plenum Press, New York
4. Guzelian AA, Katari JEB, Kadavanich AV, Banin U, Hamad K, Juban E, Alivisatos AOP, Wolters RH, Arnold CC, Heath JR (1996) Synthesis of size selected, surface passivated InP nanocrystals. J Phys Chem 100:7212
5. Birkholz M (2006) Thin film analysis by X-ray scattering. Wiley-VCH Verlag GmbH and Co. KGaA
6. Suo X, Guo X, Li W, Planche M-P, Bolot R, Liao H, Coddet C (2012) Preparation and characterization of magnesium coating deposited by cold spraying. J Mater Process Tech 212:100–105
7. Li Y-Q, Kang Y, Xiao H-M, Mei S-G, Zhang G-L, Fu S-Y (2011) Preparation and characterization of transparent Al doped ZnO/epoxy composite as thermal-insulating coating. Compos B 42:2176–2180
8. Recco AAC, Tschiptschin AP (2012) Structural and mechanical characterization of duplex multilayer coatings deposited onto H13 tool steel. J Mater Res Technol 1(3):182–188
9. Raoufi D, Kiasatpour A, Fallah HR, Rozatian ASH (2007) Surface characterization and microstructure of ITO thin films at different annealing temperatures. Appl Surf Sci 253:9085–9090
10. Iwai H, Hammond JS, Tanuma S (2009) Recent status of thin film analyses by XPS. J Surf Anal 15(3):264–270
11. Chalker PR, Bull SJ, Rickerby DS (1991) A review of the methods for the evaluation of coating substrate adhesion. Mater Sci Eng A140:583–5929
12. Yacobi BG, Holt DB, Kazmerski LL (1994) Microanalysis of solids, In: Yacobi BG, Holt DB (Eds), Scanning electron microscopy. Plenum Press, New York/London, pp. 25–26
13. Chu PK, Li L (2006) Characterization of amorphous and nanocrystalline carbon films. Mater Chem Phys 96:253–277
14. Vernon-Parry KD (2000) Scanning electron microscopy: an introduction, Ill-Vs Rev 13(4)
15. Dan JP, Boving HJ, Hintermann HE (1993) Hard coatings, J De Physique IV 111:933–941
16. Fahlman BD (2011) Materials chemistry, 2nd edn. Springer, New York, pp. 585–667
17. Oliver WC, Pharr GM (2004) Measurement of hardness and elastic modulus by instrumented indentation: advances in understanding and refinements to methodology. J Mater Res 19:3–20
18. Pharr GM (1998) Measurement of mechanical properties by ultra-low load indentation. Mater Sci Eng A 253:151–159
19. Azim MdA (1994) Deposition and comparative wear study of thin film coatings. A thesis of M Eng submitted to Dublin City University. pp 47–52
20. Werner HW, Garten RPH (1984) A comparative study of methods for thin-film and surface analysis. Rep Prog Phys 47:221–344

Chapter 7
Conclusions and Future Scope

Abstract Surface alloying is a group of surface engineering family. Surface alloying is an important method used for many automobile components, food processing, and nuclear industries. Chapter 1 has covered the basics of surface alloying. Theme of the other chapters is related to the important surface alloying methods, like carburizing, nitriding, chromizing, and duplex treatment. One chapter is specifically devoted to the characterization of surface layers. Future scope of the surface alloying is promising. Some of the interesting ongoing research activities in the following fields are briefly outlined: (i) severe plastic deformation of surface on surface alloying of ferrous and nonferrous metals/alloys, (ii) laser surface alloying, (iii) friction-stir surface alloying, and (iv) lasma source ion implantation, etc.

Keywords SMAT · Surface alloying · Friction-stir alloying · Ion implantation

7.1 Conclusions

- Surface alloying is a widely used method in industries to improve the surface properties of metals/alloys. In many industrial applications, like gears, piston rods/rings, crankshaft, etc., the requirements from the surface of components are different than the requirements from the core. The fatigue, tribological, and/or anti-corrosion properties of less expensive grades of alloys can be improved by using the surface alloying treatments such as carburizing, nitriding, chromizing, etc. Basics associated with the surface alloying have been discussed in Chap. 1. Improvements in the mechanical properties of the surface due to the surface alloying are directly related to the formation of new phases (e.g., compounds of alloying elements) and the development of residual macro-/micro-stresses/strains in the surface layer. The important process parameters are: temperature,

S. S. Hosmani et al., *An Introduction to Surface Alloying of Metals,*
SpringerBriefs in Manufacturing and Surface Engineering,
DOI: 10.1007/978-81-322-1889-0_7, © The Author(s) 2014

time, and chemical potential of the species in atmosphere surrounding the workpiece.

- Nitriding of Fe-Cr and Fe-V alloys showed two types of nitride precipitation morphologies in the diffusion zone: (i) continuous (appears "bright" under optical microscope) and (ii) discontinuous (appears "dark" under optical microscope). Discontinuously precipitated region has lamellar morphology and their formation depends on the concentration of alloying elements. This precipitation morphology is observed in the nitrided Fe-Cr and Fe-V alloys containing the alloying elements concentration more than about 2 wt%. However, these precipitation morphologies are not reported in the literature for the nitrided Fe-Ti and Fe-Al alloys. Nitrided Fe-Me (Me = Cr/V/Al/Ti) alloys show the presence of "excess nitrogen" in the nitrided zone. It has been realized that the total amount of excess nitrogen is the combination of (i) mobile and (ii) immobile excess nitrogen. Mobile excess nitrogen is responsible for more nitriding depth than the expected value. The compound layer formed on the nitrided Fe-7 wt%Cr and Fe-4 wt%V alloys was not just a pure γ'-Fe$_4$N, but it was the combination of γ' plus nitrides of alloying elements.

- The results associated with the effect of different operating parameters of plasma nitriding, gas nitriding, and nitrocarburizing of 4,330 V steel (NiCrMoV low-alloy high-strength steel) have been mentioned in Chap. 3. In case of plasma nitrided 4,330 V alloy, different geometries of the specimen have showed different kinetics of nitriding. White-layer penetration and the formation of iron nitride network form irrespective of the type of processing technique, i.e., gas/plasma nitriding and nitrocarburizing. It has been observed that at higher temperatures, white-layer penetration and the density of iron nitride network at corners are considerably higher compared to that of at lower temperatures. However, such behavior has not observed in the plasma-nitrided specimens. By controlling plasma-nitriding temperature and the N$_2$:H$_2$ gas ratio, the monophase compound layer can be obtained on the specimen surface.

- Surface alloying of materials by plasma nitriding using a glow discharge has become an important environmentally benign surface modification process to obtain improved hardness, wear, and corrosion resistance. Unlike the conventional gas-nitriding process, plasma used in the nitriding process enhances the nitriding kinetics even at low temperatures. The kinetics of nitriding, hardness profiles, and mechanism of nitriding in type 316 stainless steel, titanium-modified austenitic stainless steel (D-9 alloy), and chrome-plated type 316 stainless steel have been discussed in Chap. 4.

- Both type 316 stainless steel and D-9 alloy indicated a similar nitriding behavior by surface alloying by direct current plasma nitriding. The plasma-nitrided layers contained an iron-rich surface zone consisting of mostly iron nitrides followed by a subsurface zone of CrN precipitates. Both the nitrided materials showed a maximum surface hardness of about 1,000 HV and the hardness-depth profiles showed a sharp interface between the nitride case and the matrix. Since these steels contain Cr $\geq$ 15 wt% and these may be considered as a system in which a strong nitrogen–solute interaction prevails. Such a strong interaction results in the formation of a uniformly hard subsurface of CrN which advances progressively into the core causing a sharp boundary between the nitrided layer and the bulk. From the concentration-depth profiles of nitrogen in D-9 steel, the diffusion coefficients of nitrogen in the nitrided layer and the activation energy for the diffusion of nitrogen were calculated using the rate equation of internal nitriding model.

- In contrast to the nitriding behavior of austenitic stainless steel, plasma-nitrided chrome-plated austenitic stainless steel showed that the coating is a mixture of polycrystalline Cr_2N and Cr. The conversion of Cr into CrN occurs by nitrogen diffusion, the slow kinetics of which necessitates long process duration. A mechanism for the nitrogen diffusion in the Cr coating has been proposed on the basis of direct and indirect transfer of nitrogen into Cr coating. Indirect transfer of nitrogen could possibly occur due to electron–gas interactions through the decomposition of metastable CrN to form Cr_2N. The activation energy for nitrogen diffusion in chromium was found to be 131.4 kJ/mole. This value is much higher than the activation energy for diffusion of nitrogen in the nitrided layer of type 316 austenitic stainless steel (69.4 kJ/mole). Surface hardness of the coating is about 950 VHN. Hardness has been found to decrease gradually with depth. The abrasive wear rate of the coating was nearly an order superior to that for Cr coating and the dimensional changes of the chrome-nitrided component were less than 0.01 %.

- The effect of different types of surface treatment, like carburizing, chromizing, and duplex surface treatment (carburizing-then-chromizing and chromizing-then-carburizing), on microstructure and hardness of mild steel has been discussed in Chap. 5. Chromium carbide layer has formed at the surface of chromized mild steel. "Carbon-depleted region" is formed between chromium carbide layer and nontreated core which is responsible for undesirable hardness-depth profile across the cross-section of specimen, i.e., abrupt decrease and then, again increase in hardness with increasing depth from the surface. Similar behavior is observed for the carburized-then-chromized specimen. However, in case of the chromized-then-carburized specimen, carbon-depleted region (caused by former chromizing process) is filled with carbon during carburizing. Therefore, the continuous decrease in the hardness with depth is observed within the chromized-then-carburized surface layer of the specimen. Microstructure of the cross-section near to the surface of specimen consists of chromium carbide layer followed by pearlite colonies.

7.2 Future Scope

7.2.1 Effect of Surface Mechanical Attrition Treatment on Surface Alloying of Metals/Alloys

Surface mechanical attrition treatment (SMAT) is the process in which grain refinement of coarse grains occurs to accommodate the plastic deformation induced by the impacting hard balls (diameter in range of 3–8 mm) with velocity of about 1–20 m/s (velocity of the balls is due to the vibration–frequency in the range of 50–20 kHz) [1]. Due to the variation in strain and strain rate from the treated top surface to the deep matrix, grain size of the surface layer varies from a few nanometers (in the top surface layer) to several micrometers. Surface hardness of the SMATed iron is more than twice compared to coarse grain counterpart. This surface hardness does not reduce even after annealing for 1 h at 593 K [1].

Li et al. conducted study on AISI 4,140 steel to understand the effect of SMAT on plasma-nitriding behavior [2]. Low temperature (below 490 °C) plasma nitriding of SMATed specimen has more thickness of white-layer than the non-SMATed specimen. However, white-layer thickness is comparable to non-SMATed specimen at higher temperatures. This could be due to the recrystallization of SMATed layer at elevated temperatures. Another study on pure iron [3] showed no iron nitride formation on the surface of non-SMATed specimen after gas nitriding below 400 °C. However, iron nitrides are formed on the SMATed iron specimen even at 300 °C.

Toughness of the SMATed 20CrMo alloy has been improved after plasma nitriding [4]. Cracks are formed surrounding the microhardness indentation for the non-SMATed and plasma-nitrided 20CrMo specimens. However, such cracks are not formed for the SMATed and plasma-nitrided 20CrMo specimens [4].

In case of chromizing of low carbon steel, the SMATed specimen shows that the chromium compounds formation temperature is much lower (400 °C) and the amount of chromium carbides is higher than those in the coarse-grained counterpart [5]. The enhanced chromizing kinetics originates from the numerous grain boundaries with an excess stored energy in the nano-structured surface layer due to severe plastic deformation during the SMAT [5].

7.2.2 Laser Surface Alloying

Titanium and its alloys are widely used as a biomaterial because of their good corrosion resistance, high specific strength, and biocompatibility [6–8]. However, the applications of pure titanium are sometimes limited due to its low surface hardness and poor wear resistance. Laser as a source of monochromatic and coherent radiation has a wide ranging applications in materials processing [8]. Laser surface remelting of titanium in nitrogen-containing environment is popularly known as laser gas nitriding (LGN) [8, 9]. The major advantages associated with laser-assisted surface treatment over conventional diffusion-based surface treatment are as follows [8]: (i) ability to deliver a large power/energy density (103–105 W/cm^2), (ii) high heating/cooling rate (103–105 K/s), and (iii) solidification velocities (1–30 m/s). Surface microhardness of the laser-nitrided Ti-6Al-4V substrate can be improved to about 900 VHN as compared to 260 VHN of as-received substrate [8].

The alloying additions used in the laser surface alloying process are usually metal (e.g., Co, Cr, Mn, Nb, Ni, Mo, V, W), alloys (e.g., superalloys, stellits), ceramics (e.g., carbides, nitrides, borides). Surface alloying of aluminum by using transition metals, like Ni, Cr, Mo, W, provides good corrosion resistance [10, 11]. However, these elements have low solubility in α-Al under equilibrium conditions. In this regards, attempts have been made to increase the solubility of such elements in α-Al by utilizing high rate of resolidification caused by laser treatment [11].

7.2.3 Friction-Stir Surface Alloying

Friction-stir processing is an emerging surface-engineering technology. Friction-stir processing has been applied successfully to Al, Cu, Fe, and Ni-based alloys where improvements in properties have been observed. This technique can eliminate casting defects and refine microstructures and thereby, the various properties (e.g., strength, fatigue) of the metals/alloys can be improved [12–15]. Friction-stir processing technology involves plunging a rapidly rotating, nonconsumable tool, comprising a profiled pin and larger diameter shoulder, into the surface and then traversing the tool across the surface [12]. Frictional heating and extreme deformation occurs causing plasticized material (constrained by the shoulder) to flow around the tool and consolidate in the tool's wake [14, 15]. In aluminum alloys, friction-stir processing zones can be produced to depths of 0.5–50 mm, with a gradual transition from a fine-grained, thermodynamically worked microstructure to the underlying original microstructure [14].

It is known that aluminum matrix composites (AMCs) reinforced with ceramic particles exhibit good combination of properties, like strength, elastic modulus, wear resistance, creep resistance, and fatigue resistance. Therefore, AMCs are promising structural materials for aerospace and automobile industries. However, these composites suffer a great loss in ductility and toughness due to the incorporation of nondeformable ceramic reinforcements which limits their wide applications [16, 17]. However, wear-resistant hard surface and tougher core is a good combination to enhance the life of the components. Such combination could be achieved by reinforcing the surface layer of the ductile components with ceramic particles. Such modified surface layer is known as surface metal matrix composites (SMMCs) [17]. There are several surface modification techniques, like high energy laser beam, plasma spraying, and electron beam irradiation, to fabricate SMMCs. However, these techniques involve the formation of liquid phase due to high process temperature, and therefore, the formation of some detrimental phases in the surface layer could occur during surface alloying and/or during solidification. To overcome this issue, solid-state surface processing, like friction-stir surface alloying, is a promising technique [17, 18]. The incorporation of SiC particles [18, 19] and TiC [17] particles on the surface of aluminum alloys were successful and these particles bonded well with the matrix.

7.2.4 Plasma Source Ion Implantation

Plasma source ion implantation (PSII) is a new cost-effective surface modification process compared to that of ion implantation. In PSII, substrates to be modified are placed directly in a plasma source and then pulse biased to a high negative potential. Ion implantation is restricted by its line sight nature, while in PSII, a plasma sheath forms around the substrate and the entire surface of the substrate is bombarded by ions [20]. It uses a pulsed power supply with a maximum voltage of 25 kV and a current of 10 A with varying frequency from 10 to 5 kHz. In order to obtain the full ion energy at the substrate surface, the pressure must be kept sufficiently low (<0.5 Pa) in order to avoid ion-neutral collisions in the sheath. Thus, low-pressure plasmas have to be employed. As a further advantage of PSII, it can be combined with the thin film deposition system and the method is popularly known as plasma immersion ion implantation-assisted deposition (PIID). Like ion beam-assisted deposition, PIID may be used with nonreactive or reactive gases, enabling the formation of compound films.

At present, PSII techniques have not yet been established in broad fields of industrial production although several applications appear to be very promising. PSII has to compete with a number of techniques which are well established and have been developed through many years. A comparison of plasma-nitriding and PSII-treated stainless steels shows the superiority of PSII over plasma nitriding, however, at higher capital cost. The main advantage of PSII is the availability of higher ion energies, with a number of potential beneficial consequences for practical applications. For example, PSII of nitriding of austenitic stainless steel is of particular interest. It is possible to limit surface nitride formation by PSII due to relatively large depth of implantation. In such cases the so-called expanded austenite is formed, which has the austenite structure with lattice expansion of about 7 %. By diffusion through this layer, nitrogen is continuously added to the interface with the underlying bulk, thus extending the thickness of the nitrided layer without having the precipitates of CrN [21]. No or merely a very small fraction of CrN can be detected. Compared to the untreated surfaces, the wear rates are drastically reduced, by about three orders of magnitude and the corrosion resistance is not deteriorated in the case of stainless steel. The potential use of ion implantation in tribology has been recognized long. Nevertheless, cost of the process has impeded a major breakthrough, although ion implantation is routinely applied to the hardening of surgical prostheses [22]. Therefore, with its prospective for cost reduction, PBII is a promising new technique in the surface modification of biomaterials.

References

1. Lu K, Lu J (2004) Mater Sci Eng A375–A377:38–45
2. Li Y, Wang L, Zhang D, Shen L (2010) Appl Surf Sci 257:979–984
3. Tong WP, He CS, He JC, Zuo L, Tao NR, Wang ZB (2006) Appl Phys Lett 89:021918-021918-3
4. Sun J, Tong WP, Zhang H, Zuo L, Wang ZB (2012) Surf Coat Technol 213:247–252
5. Wang ZB, Lu J, Lu K (2005) Acta Mater 53:2081
6. Meletis EI, Cooper CV, Marchev K (1999) Surf Coat Technol 113:201
7. Probst J, Gbureck U, Thull R (2001) Surf Coat Technol 148:226
8. Biswas A, Maity TK, Chatterjee UK, Manna I, Li L, Majumdar JD (2006) Trends Biomater Artif Organs 20:68–71
9. Mordike BL (1986) In: Draper CW, Mazzoldio P (eds) Laser surface treatment of metal. Martinus Nijhoff Publishers, Dordrecht, NATO ASI Series, 115:389–412. doi:10.1007/978-94-009-4468-8_36, ISBN: 978-94-009-4468-8
10. Das DK (1994) Surf Coat Technol 64:11
11. Watkins KG, McMohon MA, Steen WM (1997) Mater Sci Eng A231:55–61
12. Rhodes CG, Mahoney MW, Bingel WH, Spurling RA, Bampton CC (1997) Scr Metall 36:69
13. Mahoney MW, Rhodes CG, Flintoff JG, Spurling RA, Bingel WH (1998) Metall Mater Trans 29A:1955
14. Mishra R, Mahoney M, McFadden S, Mara N, Mukheerjee A (2000) Scr Mater 42:163–168
15. Mahoney M, Mishra R, Nelson T, Flintoff J, Islamgaliev R, Hovansky Y (2001) In: TMS proceedings, friction stir welding and processing, pp 183–194, 4–8 Nov 2001
16. Miracle DB (2005) Metal matrix composites—from science to technological significance. Compos Sci Technol 65:2526–2540
17. Thangarasu A, Murugan N, Dinaharan I, Vijay SJ (2012) Sadhana 37(5):579–586
18. Mishra RS, Ma ZY, Charit I (2003) Friction stir processing: a novel technique for fabrication of surface composite. Mater Sci Eng A341:307–310
19. Mahmoud ERI, Ikeuchi K, Takahashi M (2008) Fabrication of SiC particle reinforced composite on aluminium surface by friction stir processing. Sci Technol Weld Joining 13:607–618
20. Baba K, Hatada R (2004). In: Patil DS et al (eds) Proceedings of DAE-BRNS workshop on plasma surface engineering. Allied Publishers Ltd, Mumbai, pp 28–44
21. Mukherjee S (2004). In: Patil DS et al (eds) Proceedings of DAE-BRNS workshop on plasma surface engineering. Allied Publishers Ltd, Mumbai, pp 181–199
22. Sioshansi P, Tobin EJ (1996) Surf Coat Technol 83:175

About the Book

An Introduction to Surface Alloying of Metals aims to serve as a primer to the basic aspects of surface alloying of metals. The book serves to elucidate fundamentals of surface modification and their engineering applications. The book starts with basics of surface alloying and goes on to cover key surface alloying methods, such as carburizing, nitriding, chromizing, duplex treatment, and the characterization of surface layers. The book is useful to students at both the undergraduate and graduate levels, and also to researchers and practitioners looking for a quick introduction to surface alloying.

S. S. Hosmani et al., *An Introduction to Surface Alloying of Metals*,
SpringerBriefs in Manufacturing and Surface Engineering,
DOI: 10.1007/978-81-322-1889-0, © The Author(s) 2014

133